CW01272105

LIVING WITH LYNX

Jonny Hanson is an environmental social scientist at Queen's University Belfast and an award-winning social entrepreneur who set up and managed Northern Ireland's first community-owned farm. Raised between Malawi, Africa, and Monaghan, Ireland, he has a PhD in snow leopard conservation from the University of Cambridge and is an Affiliate of the Snow Leopard Conservancy.

www.jonnyhanson.com

@jonnyhhanson

LIVING WITH LYNX

Sharing Landscapes with Big Cats, Wolves and Bears

JONNY HANSON

PELAGIC PUBLISHING

First published in 2025 by
Pelagic Publishing
20–22 Wenlock Road
London N1 7GU

www.pelagicpublishing.com

Living with Lynx: Sharing Landscapes with Big Cats, Wolves and Bears

Copyright © 2025 Jonny Hanson

The moral rights of the author have been asserted by him in accordance with the Copyright, Designs and Patents Act 1988.

All rights reserved. Apart from short excerpts for use in research or for reviews, no part of this document may be printed or reproduced, stored in a retrieval system, or transmitted in any form or by any means, electronic, mechanical, photocopying, recording, now known or hereafter invented or otherwise without prior permission from the publisher.

https://doi.org/10.53061/ALHK5596

A CIP record for this book is available from the British Library

ISBN 978-1-78427-495-5 Hbk
ISBN 978-1-78427-496-2 ePub
ISBN 978-1-78427-497-9 PDF
ISBN 978-1-78427-548-8 Audio

Cover image: Eurasian lynx *Lynx lynx* © Mark Hamblin/scotlandbigpicture.com

Typeset in Minion Pro by S4Carlisle Publishing Services, Chennai, India

To my grandfather, Edward Hegarty,
who first taught me to farm

To the nation of Malawi,
who first introduced me to the wild

To all those who live with chronic illness,
and to all those who care for them

Contents

	Acknowledgements	viii
	Preface	x
PAST		**1**
1	The Lay of the Land	3
2	Our Inner Landscapes	19
3	Where the Wild Things Aren't	33
4	Comeback Kids	47
5	Missing Links	63
PRESENT		**79**
6	The Lynx Will Lie Down with the Lamb	81
7	Money Talks	96
8	A Deadly Game	113
9	Animal Spirits	129
10	Common Ground	145
FUTURE		**161**
11	The Call of the Wild	163
12	Political Animals	180
13	Home Sweet Home	197
14	For the Love of Wisdom	213
15	Reconciliation	231
	Notes	247
	Index	264

Acknowledgements

Many people have made this book possible in many ways.

Thanks to everyone at the Nuffield Farming Scholarships Trust for providing the space, time and resources to think about the future of farming nationally and globally, and to consider this particular topic in depth through study and travel. Thanks also to all my fellow 2023 Nuffield Farming Scholars – across the UK and the world – who have provided me with multiple conversations and perspectives on this issue. I am grateful to the Thomas Henry Foundation for their generous financial support.

Thanks to Nigel, David, Sarah and the rest of the team at Pelagic for helping to turn my idea for this book into a reality.

My research assistant Jack Crone, on placement from Queen's University Belfast, helped turn chapter outlines into thoroughly researched and referenced chapter plans: thank you.

Thanks to all the individuals and organisations who allowed me to visit/interview/stay with them while researching this book, and the accompanying technical report:

In Britain and Ireland: Grace Reid of the National Sheep Association; the National Farmers' Union (NFU); Peter Cairns of Lynx to Scotland; David Smyth of Rewilding Ireland; NFU Scotland; the Ulster Farmers' Union; Alastair Driver of Rewilding Britain; Hedd Pugh of NFU Cymru; Eddie Punch of the Irish Cattle and Sheep Farmers' Association; the Irish Farmers' Association; and Hugh Marcus.

In Switzerland: Nikolaus Heinzer and Elisa Frank of the University of Zurich; Daniel Mettler of AGRIDEA; Thomas Jaeggi of the Swiss Farmers' Union; Christian Stauffer of KORA; Sandro Michael of the Bündner Bauernverband; Stefan Geissmann of Plantahof; Marcel Zueger of Pro Valladas; Sarah Zippert and her extended family; an anonymous Swiss conservationist; and an anonymous Swiss sheep farmer.

In the Netherlands: the LTO (Dutch Farmers' Union); Daniel Verissimo, Sophie Monsarrat and Jelle Harms of Rewilding Europe; Glenn Lelieveld of Zoogdiervereniging; Martin Drenthen of WildlifeNL and Radboud University; Richard and Stefana van de Wetering; and Carina Bakker-van de Beek and her extended family. A special thanks to my Afro-Dutch family: Emile, Maurice, Sara, Jorien, Noah and Taru.

In the USA: Eric Odell of Colorado Parks and Wildlife; an anonymous North Park rancher; Joanna Lambert and Alma Sanchez at the University of Colorado – Boulder; Tom Kourlis; Don Hunter of the Rocky Mountain Cat Conservancy; Mireille Gonzalez at Colorado State University; Ben Guillon of Conservation Investment Management and the Wildlife Friendly Enterprise Network; Jason Fearneyhough and family; Jim Magagna of the Wyoming Stock Growers Association; Bob Budd of Wyoming Wildlife and Natural Resources Trust; Ralph Brokaw; Shaun Sims and family; Dan Thompson, Mark Aughton and colleagues of the Wyoming Game and Fish Department; Jaden Bales of the Wyoming Wildlife Federation; Joe Kondelis of the American Bear Foundation; Ellery Vincent, Malou Anderson Ramirez and family at Anderson Ranch; Becky Weed and Dave Tyler at Thirteen Mile Lamb and Wool; Mitch Doherty of Vital Ground; Christopher Preston at the University of Montana – Missoula; and Quentin Martins at True Wild in Sonoma. A special thanks to my fellow members of Team Snow Leopard at the Snow Leopard Conservancy, also in Sonoma: Rodney Jackson, Darla Hillard, Ashleigh Lutz-Nelson, Kayley Bateman and Astrid Stevenson.

Elsewhere: Justine Alexander, Rob Howe and Clement Hodgkinson in France; Niall Curley of COPA-COGECA in Belgium.

Thanks to all those who read drafts of chapters: Katie McGaughey, Sinead O'Sullivan, Peadar Brehony, Brian Peniston, Peter Damerell, Don Hunter, Pete Edgar, Matt Williams, Maurice Schutgens, Drew Mikhail, Sean Jones, Darla Hillard and Emile Schutgens.

Thank you to my parents who brought my siblings and me up between worlds. For a time, I did not know which one I belonged to. But in the end, I came to belong to all of them, and they to me. It made me.

Thanks to my children, Joshua, Bethany and Sophia, who will inherit this beautiful and broken world: someday the baton will be your generation's to carry.

And last but not least, to my wife Paula: thank you for shouldering the responsibility of home and children when I've been away on fieldtrips. Through thick and thin.

Preface

This is the story of my rewilding coexistence adventure.

To write it, I travelled 2,115 km on 45 trains, flew 21,280 km on 12 planes and drove over 4,400 km by car. I interviewed and visited over 50 people and organisations across seven countries. I journeyed back in time, across the world and into the future.

I also delved deeply into the human psyche, including my own.

I invite you to join me on this adventure, as we delve deeply into what is, arguably, the most controversial environmental issue of twenty-first-century Britain or Ireland: whether we could and should learn to live with lynx, wolves and bears again.

Too often, decisions about topics like this are discussed, debated and made only by people with PhDs in ivory towers. In contrast, this book is for anyone to discuss down the pub. It's for people and their mates to chat about over coffee. It's for everyone. Because each one of us, across these islands, deserves the opportunity to be informed about the nuanced dimensions of this complicated issue. And to feel and think of it what we will.

It's also a topic I have mixed feelings about. This is my attempt to explore these feelings with honesty and humour. I suspect that the majority of British and Irish citizens also have mixed feelings about it, in contrast to vocal minorities opposed or in favour.

But whether your feelings are fixed or mixed, this rewilding coexistence story is for you. And like all good stories, it begins like this: once upon a time, in a land far, far away…

PAST

CHAPTER 1

The Lay of the Land

The wind howled. I glanced at the moody sky, and then at the bleak landscape of this County Antrim hillside. Above me, dark clouds were roiled by a brutal, driving wind from the south-west. I felt the first specks of rain on my cheek. Below me, to the east, I could see the colours of autumn creeping through the woods of Glenariff, the Queen of the Glens. Orange for the beeches; yellow for the larch; and evergreen for the tall and sombre plantation conifers in their serried ranks. Far down the valley, I glimpsed the small village of Waterfoot, spotlighted in a distant beam of sunshine, pushing back against the overwhelming greyness of the day. The Irish Sea beyond it looked angry, as if fighting with the wind. I could make out the white-cap waves from here, gnashing their teeth against the shore.

Around me were the rolling hills of the Antrim plateau. A dull palette of browns and greys comprised this October day, as I skirted the edge of Dungonnell reservoir towards the start of the Garron plateau. Gusts of wind skated over the dark water. I passed some sheep huddled in the lee of a rocky outcrop. It was a stark landscape on an equally stark day. But in its starkness it was stunning. And in its starkness it was home. Buffeted by the wind at my back, I pulled my coat tighter. I kept walking.

My walk was my usual Monday morning stroll, a habit I had kept for several years as a form of self-care. Not only did I find it the easiest way to start the working week, but also the most productive. Unstructured time in nature can improve both well-being and productivity.[1] Thoughts spring unbidden to my mind as I casually, and often unconsciously, connect previously disparate pieces of information. Sitting down to formally start working later on, and for the rest of the week,

I often found myself merely implementing the solutions that had formed in my head first-thing on Monday.

But the thoughts gusting around my mind that morning looked backwards, not forwards. That's because as I kept walking along the northern edge of the dam, I was walking back in time. These lands once formed part of the vast estates of the Earls of Antrim, at their height in the seventeenth century totalling over 330,000 acres.[2] From their seat in Glenarm, like many of the Anglo-Irish aristocracy, they were plugged into the corridors of power in London and elsewhere. The imposing edifice of Garron Tower was built as a summer house by Frances Anne Vane Tempest, Lady Londonderry, in 1850, later becoming a school in 1951.[3] The local hotel in nearby Carnlough, built as a coaching inn by Lady Londonderry in 1848, was even inherited by Winston Churchill in 1921.[4]

It was here, on the Garron plateau, in this remote corner of the then Kingdom of Ireland, in 1712, that the last Irish wolf (*Canis lupus lupus*) was shot.[5] Or so legend has it. In fact, as Kieran Hickey recounts in his *Wolves in Ireland: A Natural and Cultural History*, there are several places that lay claim to being the location of the last wolf in Ireland and, by default, the last large carnivore in Britain and Ireland. Hickey cites Co. Carlow in 1786 as the most likely location. It is therefore only relatively recently in historical terms, and the blink of an eye in geological terms, that these animals lived alongside us on these islands. While wolves were extirpated from Ireland in the late eighteenth century, their extinction in Britain occurred much earlier: around 1304–5 in England, also the fourteenth century in Wales, and 1684 in Scotland.[6]

Other big predators disappeared earlier. Eurasian lynx (*Lynx lynx*) were last definitively recorded in Britain around the late fifth or early sixth century CE,[7] although evidence is emerging of the species likely presence in early modern Scottish texts.[8] In Ireland, the single piece of physical evidence of their presence on the island dates from 8,875 years ago.[9] On the other hand, brown bears (*Ursus arctos*) were present in Ireland until the Bronze Age around 1000 BCE,[10] while in Britain there is some contention about whether the species also disappeared around this time or persisted through Roman times until the early medieval era.[11] There is a clear spatial pattern of large carnivores clinging on in the upland, northern and western parts of Britain and Ireland, a trend with equally clear resonance for today's reintroduction debate, as we shall see later on.

Given the mostly prehistorical timeframes of lynx and bears vanishing from Britain and Ireland, there is limited historical evidence for how people viewed or interacted with them. Bear bones from Co. Clare's Ailwee caves, for instance, show butchery marks, suggesting the species was hunted for food.[12] Wolves, though, are the exception, given their persistence until more recent times, especially in Ireland. Historical records, and cultural references, abound. While *Canis lupus lupus* may be gone from these shores, its presence lives on in our back catalogue of fairy tales, and their modern billion-dollar-blockbuster equivalents from Disney, Dreamworks and the like. What is clear throughout, though, is that through a combination of direct hunting plus competition for land, deer and livestock, humans drove these species to extinction.

Back in the Glens of Antrim, the howl of the wolf had been replaced only by the howl of the wind. A ranger from the public utility that operates the reservoir and owns the surrounding watershed drove past on his way to a pumphouse. Further along, a pair of farmers on a quad bike trundled alongside. We raised hands in greeting, instead of futile words that would be instantly borne away by the gale. Dressed head to toe in wet-weather gear, standard issue for their hard outdoor job, they were off to check their sheep, in what was just another day at the office.

I was walking through a living landscape. The surrounding hills captured and filtered the water for the reservoir, stained red by the peat it passed through. In turn, this supply provided for the domestic and industrial needs of the town of Ballymena, my current home, and much of the rest of mid Antrim. Ecosystem services like this underpin all human activity on Planet Earth. Nature provides, for free, services – like water filtration, pollination and climatic regulation – that humans cannot fully, or sometimes even partially, replicate. They are a reminder of the significance of natural places like this for human survival and flourishing. They are also a reminder of nature's hidden ecological economy – both ecology and economy come from the Greek *oikos*, meaning 'home' – at work, transforming a diverse set of inputs, via processes, into usable outputs that benefit our and other species. By far the most productive economy on Planet Earth is Planet Earth.

In economic terms, this part of the Antrim plateau was also a working landscape. Ruined cottages throughout the countryside of Britain and Ireland, but especially in their upland areas, hint at the significant rural depopulation that has occurred over the past decades and centuries. Yet people still live and work here, keeping mostly sheep on these high hillsides through summer and autumn, a form of seasonal transhumance still alive and well in the twenty-first century. Much ink has been spilled about the ecology and economy of upland sheep farming. Irrespective of that, it remains a valued part of rural and upland life, with its practitioners, like those who had just passed me on the quad, often hefted to a piece of land over generations.

A meadow pipit flitted across my path, bouncing between the heather on its lilting flight. This was a diverse landscape too. Thick layers of peat hinted at dense primordial forests and other ecosystems that had once covered these lands. Now only stands of plantation conifers dotted the landscape. So too did the numerous drainage channels, criss-crossing the area around the reservoir. More recently, a conservation partnership, between the water utility and a conservation charity, had worked to dam these channels, restoring peatland areas to a more sodden state. Flora and fauna that benefited from these waterlogged conditions were making a comeback. Like the curve-billed curlew that I had disturbed earlier, or the elusive marsh harrier that I had yet to spot.

This was a form of rewilding in action. Rewilding, as one definition would have it, involves the return of processes to landscapes, often by returning species that facilitate those processes[13] The re-dammed drainage channels around Dungonnell reservoir slowed the movement of water through the landscape, not only improving wildlife habitat but also providing downstream benefits, such as reducing soil erosion and therefore lessening the amount of silting in the reservoir itself. The idea of rewilding developed in North America in the 1980s and quickly spread around the world. Conceptually, it proved an attractive concept, an opportunity to reverse the prevailing trend of seemingly unstoppable nature losses, increasing rather than decreasing habitat and species.

In practice, it was rather contentious. Differing views on and visions for nature caused conflict between the many different social groups, or stakeholders, with an interest in the management and use of land and its inhabitants. This was especially the case when rewilding involved the return of previously extinct animals to crowded islands like Britain and Ireland. From the 1980s onwards, several raptor species were returned, including red kites, golden eagles and white-tailed sea eagles.[14] In the early 2000s, beavers were 'accidentally' returned to Scotland, quickly establishing their presence in the landscape and in people's hearts and minds.[15] Bison followed in the Weald of Kent in 2022, the first in Britain for over a thousand years.[16]

But a piece of the ecological puzzle was missing. Ecosystems exist in a state of dynamic equilibrium, where, put simply, the populations of the species in a particular layer of the food chain, be they animal or plant, are kept in check – in part – by the actions of the species in the layer above. Other factors like disease and climate also play important roles. Returning bison to woodland, for example, will change the structure of the woodland over time through their large-scale grazing, browsing and trampling of plants. But who will keep the bison in check to prevent damage to the very habitat they are meant to be conserving? The short answer is humans. For a growing number of proponents of rewilding, an alternative answer, for bison and for other large herbivores like deer across Britain and Ireland, could be large carnivores.

The potential reintroduction of large carnivores, like wolves, lynx and bears, to these islands has long been mooted by some rewilding advocates. But until fairly recently, it seemed like a far-fetched fantasy for armchair conservationists who had watched too many David Attenborough documentaries. No longer. In 2017, the Lynx UK Trust made an application for trial reintroductions of small numbers of lynx to Thetford Forest on the Norfolk–Suffolk border and Kielder Forest in Northumbria. It was declined,[17] but a consortium of rewilding organisations are considering the idea further north, through the Missing Lynx and the Lynx to Scotland projects.[18] Also in Scotland, some controversial large landholders have made no secret of their desire to return wolves to at least parts of the countryside.

The notion that large carnivores could once again roam these beloved islands I call home fills me with a range of emotions. The idea of coming face to face with a wolf or lynx on my weekly walks through the forests and hills of Co. Antrim fills me with wonder and anticipation. At the same time, the conflict, complexity and expense required to get to that point fills me with dread. Fans and critics of reintroductions will have legitimate opinions on how they could work, where they could work and, most significantly, whether they should be allowed to happen at all.

All of these thoughts swirled around my head on this blustery autumnal day, where wolves once roamed between sea and sky over the purple-headed mountain. What did the future hold for people and nature in this scenic part of the Glens of Antrim, I wondered, and for other landscapes across Britain and Ireland, both similar and different? How wild could they become? How wild should they become? To begin the process of answering these questions, I first looked inwards.

A foot in both camps

Memory took me back. Back to a crisp December morning in 2020. Magical hoarfrost cloaked the trees and the grass crunched beneath my wellies on the frozen ground. I was doing my morning rounds of the small Co. Antrim farm that I managed, only about 20 miles or so from Dungonnell reservoir. My morning routine had begun as it always did. Stumbling out of the farmhouse after breakfast and coffee, I surveyed the magnificent view. Facing east, atop a hill that sloped steeply down to the Glynn river, Jubilee Farm looked out on Larne lough. The rising sun danced on the sheet-glass-water, still as a millpond, reflecting a brilliant winter dawn.

I had headed straight to the feedshed, first checking, feeding and watering the sow and piglets in the pen close to the house. Next I headed up to the farm's one flat field, where our vegetables grew. One of the big polytunnels housed the fattening pigs over winter – the deafening din of 'hangry' pigs had started from here the moment they heard the feedshed door open. The other large polytunnel was two-thirds winter vegetables, with the remainder given over to the fattening goat kids and the pregnant nannies, about halfway through their gestation.

In a one-acre paddock next to the market garden, we kept the billy goat through the winter, along with a castrated male for company. We also kept our pair of geese with them. In part this was because there's an old theory that keeping geese and ducks with goats and sheep can reduce the incidence of parasites in the pasture and the stock, the birds disrupting the lifecycle of the liver fluke by eating the snails that host them. But mainly, it was because, in with the goats, and close to the market garden and farmhouse, with its constant bustle of people, dogs and vehicles, the geese would be safe.

I headed past the third, smaller polytunnel to break the ice on the water troughs in the paddock and to throw both geese and goats a handful of cereal mix. Headcounts morning and afternoon were standard operating procedure, as they are on any livestock farm. I counted off two goats. I counted off one goose, waiting impatiently to be fed. I went to the small ark to count off the second goose. It was gone.

In its place was a pile of white feathers where once our magnificent gander had been. I followed the trail of blood to the fenceline, where it disappeared across our neighbour's land, probably ending up in the wooded brae that filled the steep river valley below. A fox had struck in the night, or in the half-light of pre-dawn. I cursed the insolent thief.

Over the previous years of setting up and running Northern Ireland's first community-owned farm, I had lost livestock to accident, illness, weather and contaminated feed. But I had never lost any to a predator. The conservationist in me knew that the fox was merely playing the role it had in ecosystems across the world for millions of years, scavenging and hunting opportunistically. Our gander was an easy snack at a lean time of year. The farmer in me was furious. Not only because I'd raised and cared for these geese, remembering how the gander had half-escaped from his box in the boot of the car on the journey back from the seller, fixing me with a beady eye in the rear-view mirror, as the kids laughed uproariously. But also because I had high hopes that this pair would produce offspring in the coming spring that we could raise and sell the following Christmas. With a foot in both camps, I felt torn.

I have always loved farming. Some of my earliest memories involve time spent with my grandfather on his Co. Derry/Londonderry farm, around land, livestock and machinery. As a child of the nineties, growing up in rural Co. Monaghan, I spent much time on the farms of school friends, most of whose families farmed, full-time or part-time. A precocious and avid reader, I recall, at the age of ten and already a subscriber to several farming periodicals, writing to inform one poultry columnist that he had misidentified a breed of cattle on a visit to the Netherlands. They were not, in fact, Belted Galloways but, rather, Lakenvelders, or Dutch Belted. I still have the polite letter he wrote back, acknowledging his mistake. As a teenager of the noughties, I spent most summers working with my grandfather and uncles, primarily with cattle and sheep, plus the never-ending farm chores. These experiences imprinted upon me a deep love of the domestic, and of farming as a way of living, working and being.

Farming became one of two things that I wanted to do with my life. As a postgraduate student at Queen's University Management School, I specialised in sustainable food supply chain management, becoming convinced of the strategic nature of the sector for addressing a wide range of societal challenges, from biodiversity loss and climate change to human nutrition and well-being. Not having direct access to land, I later established and managed Jubilee Farm on a 13.5 acre site in Co. Antrim. Through literal blood, sweat and tears we successfully set up the first community-supported agriculture scheme in Northern Ireland, integrating it with care farming programmes that worked with both adults with learning difficulties and with refugees and asylum seekers, in the latter aspect the first of its kind in the province. Included with both of these was the third and final component of our work: conservation.

I have always loved conservation. An early obsession with animals, including heavy consumption of my dad's *National Geographic* back catalogue, David Attenborough documentaries and *The Really Wild Show*, gave way, as my worldview expanded, to a concern for their survival. I spent much of my spare time designing zoos, as well as farm parks, by my pre-teens going into significant detail, including budgets, staffing and near-architectural-level sketches of buildings and enclosures. When I was 11, we returned to Malawi, having lived there briefly for a year when I was an infant. I spent time with school friends running half-wild through the forested hills around our home city of Blantyre, scattering baboons and bushpigs before us, and encountering much larger animals on numerous family safaris in the region. There were breeding populations of both leopard and spotted hyena in some of the forest reserves around Blantyre but they wisely kept out of the way as we came thundering through.

Conservation became the second thing that I wanted to do with my life. Long university summers were spent gaining practical husbandry experience with as many different types of wild animal as I could. First across Britain and Ireland with cats big and small, primates, emus, native wildlife, reptiles, wolves/wolf-hybrids and red kites. Then in Canada with large hoofstock, from rhinos and tapirs to giraffe and antelope. As a then undergraduate in medieval history and archaeology – another great love of mine – I briefly flirted with the idea of going down the animal behaviour and welfare route, teaching myself the science of ethology (the science of animal behaviour) and how to research it. I even managed to talk my way into running an olfactory – or smell-based - enrichment experiment with a tiger and a bunch of confiscated Pit Bull Terriers at an animal rescue centre in Northern Ireland that had previously been a safari park.

Increasingly, though, I was thinking on a bigger and bigger scale, as I recognised the need to reconcile the social, economic and environmental dimensions of life on Earth. Sustainability became the mantra that brought my diverse interests and skills together. I embarked on a PhD at the University of Cambridge, to explore snow leopard conservation and its links to rural development in Nepal. Partnering with the Snow Leopard Conservancy, and invaluably helped by a number of research assistants, my team and I talked to over 700 locals and over 400 tourists in the Everest and Annapurna regions of Nepal. Key themes that emerged included: the close link between people's attitudes to snow leopards and their attitudes to the people and methods used to conserve them; the significance of animal husbandry as a factor in reducing livestock losses to predators; and the potential and challenges of tourism and, to a lesser extent, livestock product certification schemes to offset the costs of living alongside snow leopards. Later consultancy work with the Snow Leopard Conservancy unearthed a trend common in many locations globally: people are generally more tolerant of wild cats and related livestock losses, than of wolves and their impacts.

In March 2022, after eight years of finding, financing, establishing and managing Jubilee Farm, I handed it over to a new management team. It was time to scale up all of my learning from both my conservation and my farming activities. Part of the time, I began co-delivering a community farming accelerator programme, mentoring and training ten diverse, early-stage community farming projects across Northern Ireland. But large carnivores were calling. My mind was drawn back to that early December morning, to a goose and a fox.

It was a scene that had played out countless times around the globe over the last 10,000 years of human history. It was a scene that, on one hand, threatened the livelihoods of farmers, herders and ranchers, large-scale and small-scale, the world over. On the other hand, it was a scene that also threatened the existence of numerous carnivore species through retaliatory killings, many, like the snow leopard, much more threatened than red foxes. But by now, it was clear to me that farms and farmers were key to the persistence of much wildlife, including large carnivores. It was clear to me that the momentum for reintroducing large carnivores to Britain and Ireland was building and that some form of trial project was likely to go ahead in the coming years. And it was clear to me that I had a unique blend of skills and experiences to help me take a balanced look at the issues, with a foot in both camps.

I applied to the Nuffield Farming Scholarships Trust and was awarded a 2023 Nuffield Farming Scholarship. A programme of travel and study to engage future farming leaders with the future of farming, the scholarship would give me the means to travel widely and think deeply about large carnivore reintroductions to these islands, especially how they could be reconciled with agriculture. But first, I reckoned a good place to start was to meet with a fellow farmer, one older and wiser than I in the ways of the land, to seek counsel for the road ahead. Snapping out of my reverie, I stopped and turned into the wind, my cheeks instantly numbed by the cold. I headed back down from the dam and over to the next glen to meet Hugh Marcus.

Farming forward

The biting wind had subsided a little. This was one of those 'all four seasons together' kind of days. As I crested the ridge and left Glenravel behind me, I could make out the inselberg, or volcanic plug, that is Slemish. Looming large in the landscape and the imagination, the mountain is famed for its link with St Patrick, the patron saint of Ireland, who herded livestock on it as a slave in the early fifth century CE. *How did he get on with the local wolves?*, I wondered, making a mental note to consult his *Confessio*, and other lives of the saints, for evidence of past coexistence between people and predators.

Shortly after, I pulled into Hugh's yard. A buzzard wheeled on the turbulent thermals. The neighbour's cockerel crowed. Situated on rolling upland between Glencloy and Glenarm, in the townland of Ballyvaddy, this is hard land to farm.

Hugh and I chewed the fat over a cuppa, as we have many times. Then it was off to see the livestock and the land, the first time I had been beyond his yard and walked his farm with him. We stopped to see a pair of gilts that we'd sold him the previous year. No longer at the cute and cuddly stage, there was a brief and angry demand for food, as a soil-covered snout was thrust between the bars of the gate and up against my trouser leg.

'I thought this pair were destined for the freezer?' Hugh smiled wryly, and admitted he had become rather fond of these Tamworth–Oxford Sandy and Black crosses. Plans to buy a boar from another mutual contact were afoot. We agreed on a shared admiration for pigs and their enterprising attitude to life.

Next we looked in on this year's lambs, who had just come in from the hillside to fatten overwinter on haylage and oats. 'You're still OK to keep one of these for me?' I asked him. 'Though in my case, it's definitely going in the freezer!'

We crossed the road, between the drystone walls, and began walking up the slow incline to the highest point on the farm. Hugh's great-grandfather had worked in the iron ore mines near Dungonnell reservoir. When they closed in the later nineteenth century, he had bought this 40-acre farm, paying off the mortgage at the princely sum of £5 per year. I asked Hugh what lessons current and future land management could take from that generation and era.

'Diversity was the defining feature of this farm then, and of farming in general,' he remarked, gesturing to a field where plantains poked up among the grasses, at odds with the uniform nature of the neighbouring fields. These were part of Hugh's pioneering upland trial of multi-species lays, a combination of multiple grass and plant species in a single field, creating a mini agro-ecosystem. In the summer, it had been a riot of pink and white clover. 'Diversity will also ensure resilience for the future.'

We carried on to the summit of the brief rise, our breathing a little laboured. 'What about the barriers to creating diverse farms and farming systems?' I enquired.

'I think you need to start with the agricultural colleges and research institutions,' Hugh rejoined. 'The creation and sharing of knowledge tailored to the needs of farms and farming is key. It needs to help farmers adapt their sites and practices to the realities of the landscape, of the climate, of their soils.'

Hugh had shared with me earlier how heartbreaking he had found the great snowstorm of March 2013, when he had lost a third of his sheep and half of his lambs. The bureaucratic bungling that followed added insult to injury. The bales of haylage thrown from helicopters to feed starving flocks that couldn't be reached any other way did not have the plastic wrap removed. A civil servant accompanied the deadstock lorry to collect the sheep, requiring that the sacks containing the dead lambs be opened up and every head counted.

It reminded me of my own frustration in dealing with agricultural bureaucracies, a low point in setting up and running Jubilee Farm. In the worst of several occasions, I had been sent a letter reminding me about an upcoming routine test for our pig herd. The letter had promised additional support if required, via the name and number at the bottom of the page. On ringing it, I was told that that office didn't handle pig queries so I had to ring another office. That office then told me that the relevant official now worked in a separate office entirely.

When I finally got through to there and asked for the person by name, I was informed that he hadn't worked there for three months but now worked at headquarters. Blowing steam out of both ears by this point, I was put back through to the central switchboard.

'Can you put me through to headquarters?' I asked, making a valiant effort to stay civil.

'Sorry,' was the response, 'I don't have a number for there and I can't put you through.' I slammed down the phone. Five minutes later, after I'd vented some of that head of steam, I rang the switchboard again, got put straight through to headquarters by another operator, reached the official I needed to reach, and had my query dealt with constructively in under five minutes.

I ventured to Hugh my concerns about the financial compensation schemes that often accompany large carnivore conservation projects, to offset farmers' losses, being handled like this. We shared a chuckle. Like many farms across the province, and these islands, Hugh's had not been viable in and of itself for decades. Off-farm work by both him and his wife helped pay the bills and enable them to continue living on and working this land.

As we reached the top of the farm, and looked out on the lay of the land, we could see the damp area that used to be cut for peat naturally regenerating with willow. Hugh spoke about being blown by the prevailing winds of agricultural policy, winds that were not constant but that could change direction, sometimes abruptly. Yet in truth, public investment in the public goods that are food production, climate mitigation, watershed protection and biodiversity conservation is the flipside of this public involvement in countryside management. And where public money is invested, there is a continual tension between the level of trust imparted to landowners and the level of verification required by officials, their political masters and the public who elect them and hold them to account.

From our vantage point, the full splendour of the Antrim plateau lay arrayed before us. A large area of bog blanketed a hollow in the hills. Beyond that, I could make out the ridge running parallel to the Glenarm valley where the Ulster Way ran. I fondly remembered walking large sections of this with my three young children as our daily exercise over multiple Covid-19 lockdowns. To the north, I could see the edge of the Garron plateau, where I'd just come from. With the weather improving, visibility was excellent. The mosaic of habitats and colours and gradients made my heart skip a beat, as this landscape always does. It was wild and it was domestic. It was glorious.

I came to the point of my conversation with Hugh. 'Is there a place for rewilding here?' I wondered. 'And for large carnivores in particular?'

Hugh was unsure that this was the right place for the return of its once-resident wolves, given the absence of a suitable wild prey population. 'But there are certainly areas on most farms, especially in upland areas, which don't make sense to farm,' he ventured. 'Here, farmers could be supported to include rewilding as one part of their approaches to land management.'

But he added an important caveat: 'Farming is more than just a job; it's a way of life. Rewilding needs to recognise that and work with those who have lived in these upland landscapes for generations.'

We looked out on the landscape once more. The weather was a-changing. We headed back down the hillside.

I left Hugh, a sort of Northern Irish James Rebanks, striving for a kind of Ulster Pastoral on his small patch of the province. Reflecting on our conversation, it seemed that there were opportunities for farmer-focused support that could create diverse agro-environmental systems with multiple outputs and outcomes. Or translated from technocrat: a vibrant and flourishing countryside. Furthermore, the success or failure of large carnivore reintroductions would depend, to a large extent, on people like Hugh and farms like his, eking out an existence from marginal areas but deeply invested in the places they called home. I intended to talk to plenty of people like him.

Wolves and lynxes and bears, oh my

This book is the story of my journey to actively explore the complexities of large carnivore reintroductions to Britain and Ireland. As we have already seen, issues of nature and human nature, of history and natural history, and of science and stories shape this debate profoundly. My contribution to addressing some of these issues, hopefully in an inspiring and thought-provoking manner, is a rewilding travelogue in three parts.

In section one, I venture back in time. I begin by looking inwards. I consider how our inner landscapes, from evolutionary psychology to our back catalogue of fairy tales, and at the levels of both the individual and the community, are one of the most important factors in this contentious debate. Precisely because of that, we will return to this theme again and again throughout each section and chapter of the book. Asking 'How wild is wild?', I continue by introducing various important terms that we will use throughout. These include 'wild', 'rewilding' and 'reintroduction'. I explore the history of large carnivore presence and extinctions in Britain and Ireland, as well as the history of rewilding and reintroducing here. I also introduce and explore what the term 'coexistence' means in terms of our relationship with wildlife, and the extent to which we as humans can fill the gap left by the absence of lynx, wolves and bears in our landscapes.

Section two looks outward – in the present. In the absence of actually existing large carnivore reintroductions in these islands, I travel to Europe and North America to glean insights from other projects and contexts as to *how* livestock farming and predator conservation can coexist. Across five chapters I consider five broad categories of tools that are used to manage such relationships: deterrence, finance, force, enterprise and governance. All of these are set against the magnificent backdrop of Dutch, Swiss and American landscapes, supported by an all-star cast of farmers, conservationists and officials, among others.

Looking forward is the subject of section three. Building on evidence from the past and the present, I make the case for and against large carnivore reintroductions to Britain and Ireland in four important areas. These are: ecology, politics, economics and philosophy. Making precise predictions about the future may be a fool's errand, but understanding the broad trajectory of trends shaping the world we will inhabit in five or fifty years is a wise course of action in any walk of life.

Journeys always fill me with anticipation. Of distant Scandinavian stock, with a surname that is very rare in Ireland, my family likes to think that we Hansons came over with the Vikings, only a few centuries after St Patrick. Irrespective of the historical minutiae, travelling is in my blood. I felt its intrinsic urge reaching down through the centuries and across the generations. As Moana sings, in words beautifully crafted by Lin-Manuel Miranda, 'we are descended from voyagers, who made their way across the world'. Unlike my Viking ancestors, though, I was armed only with questions, the answers to which I hoped would help build bridges between the conservation and farming livelihoods that I knew and loved.

Will the last places in Britain and Ireland to host lynx, wolves and bears be the first to have them back? Will the howl of the wolf mingle once more with the howl of the wind on the Garron plateau amidst the Glens of Antrim? Could we learn to live with these animals again? Would the howls of protest, and the conflict between people with wildly differing worldviews, that this process would unleash be worth it? Above all, whether with these animals or each other, could we find ways to coexist – 'to live… together at the same time or in the same place'?[19]

My travels to answer these questions were just beginning. Leaving Hugh's farm and the Glens of Antrim behind me, I drove home, into the now driving rain. The lay of the land was arrayed before me. I braced myself for the vagaries of the terrain and climate ahead, and journeyed on.

CHAPTER 2

Our Inner Landscapes

The XL Bully charged. Powerful muscles rippled under fawn skin as it accelerated towards us. It had all happened so quickly.

Out for a rare walk by ourselves, with the kids away at their grandparents', my wife Paula and I had spotted the dog only a few seconds before. It had appeared around a corner and disappeared momentarily into some long grass. We could hear its owners talking, as they followed behind. As we continued to walk towards it, the powerfully built dog came out of the long grass and noticed us. But there was no threat display, no warning growl or bark. It saw us. It paused. It charged. Less than two seconds had elapsed.

Time slowed. My senses sharpened. Cold fog hung in the air and flickered under the streetlights. Moss clung to the edges of the tarmac footpath, as did some chewing gum. A discarded wrapper, an old shoe thrown over the wall from a neighbouring house, a steel bar sticking out from the fence opposite: the details were seared into my mind.

Still the dog came. It ate up the ground. Five of the ten metres between us disappeared in an instant. Its skin was a thin cutaneous sheath over a Schwarzenegger-esque physique. Outsized deltoids flexed hypnotically beneath it. Steaming breath came in short bursts as the XL Bully powered forward. Only now did a low rumble of a growl rip from its throat.

Instinct initiated. An older part of my brain took over, one forged on the plains of Africa a million years before. I didn't realise it at the time but with my right arm, I firmly pushed my wife backwards. As I lunged forward towards the dog on my left leg, I drew my trusty right leg back. It had served me well on the playing fields of Africa two decades before as a solid right-back, a powerhouse centre and a 400-metre champion. It had been sculpted on the squat rack the day before and trained in the swimming pool that afternoon. Now, as my quadriceps contracted and began to bring my right foot forward with force, my goal was clear. I was going to protect myself and my tribe from this predator. I was going to break its jaw.

A second had passed. Yet I saw an eternity in it. I saw my hominid ancestors chasing lions from a kill. I saw my Viking and Irish ancestors defending their cattle from wolves. I saw the succession of fairy tales I'd read and watched as a child. I saw red.

My right foot flew forward, propelled by the full extension of my quadriceps and aimed squarely at the dog's head. I could see its engorged tongue, smell its breath, almost feel its teeth. At the last possible moment, milliseconds before my foot connected with its face, it swerved. A centimetre separated us. Twice more the beast charged us. Twice more my instincts prepared me to kill, or be killed in the attempt.

The owners laughed.

Paula screamed.

A scream of pure terror. A scream of the threat of impending doom, even death. A scream that echoed through time.

Ten years earlier, as we'd been dropping off our two-year-old son at nursery in the quiet Cambridgeshire village of Longstanton, something similar had happened. We'd parked around the corner from the nursery school. It was a bitter winter's day, wind howling through the naked trees. As Paula, Joshua and several other parents and children converged on the school gates, a very large dog appeared from nowhere and, almost immediately and at random, charged at Paula. She thrust Joshua behind her, waved her hands above her head and shouted at the dog. Other parents scrambled for cover behind parked cars and lifted their children to safety. I was less than 30 metres away in the car with our infant daughter. But just out of sight, I saw nothing. And as the brutal northeaster ripped the scream from Paula's throat and bore it away, I heard nothing.

Until they came running to the car. The dog had backed off when Paula shouted at it, seemingly intimidated. She seized the opportunity to bundle Joshua and herself into the car. She was shaking with adrenaline as she blurted out the story. I checked to make sure she and Joshua were unharmed. Then I went after the dog.

It wasn't so much a desire for revenge that made me do this. From the car, I could see the dog slinking off, thankfully in the opposite direction to the school. I recognised it as a Bullmastiff, capable of serious damage to an adult, let alone a child. But there were other parents and children approaching the school from across the village. Many of these children were unaccompanied. My protective instinct kicked in.

I was going to protect my tribe from this predator. I just wasn't sure how.

I quickly searched in the car boot for a weapon. Anything. Finding nothing, I still ran after the dog, glancing everywhere for something to use. As I rounded the corner, onto the main street, I still wasn't sure how I was going to do this. Instinct compelled me onwards.

I saw the dog dart into a garden and remain there. I waited for a few minutes and then, taking a note of the house number, I beat a hasty retreat back to the car and then home with my family. It took a bit of convincing, but I eventually persuaded the police to investigate the issue. Statements were taken. Witnesses were canvassed. And a successful prosecution of the dog's owner was later carried out.

Mercifully, our son was largely unaffected by the incident. He was too young to appreciate what had happened, or had nearly happened. But Paula was traumatised. She had grown up around dogs, as I had, and loved them dearly, as did I. Yet something changed in her that day. A suspicion of strange dogs that never went away. A trauma that reared its ugly head again when we faced down the XL Bully.

Almost immediately after the Longstanton incident, the most curious thing started to happen to me. Despite not seeing or hearing the attack directly, I began to experience similar symptoms of trauma to Paula. Even though we would later get dogs of our own, and continue to enjoy time with those of our family and friends, something had also changed in me. When Paula tensed as we walked past a strange dog on the street, I tensed. As her nervous system prepared to freeze, mine prepared to fight. And this also happened when I encountered unfamiliar dogs on my own. A suspicion that has never gone away.

I was experiencing something called secondary trauma. Secondary trauma is trauma from exposure to someone else's trauma.[1] Common in caring professions, especially counselling and psychotherapy,[2] it tends to occur quickly after seeing or hearing a single traumatic incident from another. As I listened to and comforted Paula after the Bullmastiff attack, some of her trauma passed to me, as did a sense of guilt at not seeing, hearing or responding to the incident in real time.

Ten years later, when the XL Bully charged us, my response was immediate yet timeless, simple yet complex. It was layered with evolutionary instinct, with historical precedent and with cultural narrative. It was also layered with my and Paula's personal trauma. A trauma that affected our interaction with the world around us, individually and together. Our inner and outer landscapes had collided.

Primal feelings like these shape our perceptions of large carnivores. Their place in the mountains of our minds is mediated by evolution, history and culture, as well as how these feelings and beliefs are shared with others. This holds true even if we have never seen a wolf, lynx or bear in captivity, never mind encountered one in the wild. In this area, as in every other aspect of human life, we carry the weight of millions of years of human evolution, and thousands of years of human history, within us.

Understanding how and why we relate to large carnivores on this basis is therefore essential to understanding how and why we relate to their potential return to these islands. In this chapter, we'll first consider our evolutionary psychology, as well as our back catalogue of fairy tales. We'll then explore how this relates to modern psychology, including in relation to conservation, rewilding and reintroducing. We'll also explore a psychological model that explains how we can feel numerous, competing and even contradictory feelings about returning these predators to Britain and Ireland. I certainly feel like that – it's why I'm writing this book.

Our carnivore coexistence journey together takes us inward.

Primal feelings

Something lurked in the dark. The light from the snug living room cast a warm glow into the hallway. But light doesn't travel round corners. Round the corner, it was pitch black. Round the corner, something waited.

I spent my primary school years in Ballybay, Co. Monaghan. On the Carrickmacross road was one of the town's two Presbyterian manses. A rambling Edwardian house, its lands sprawled over ten acres on the edge of Ballybay. I ran wild through its fields, woodland and orchard. There was even a byre for two cows in the backyard.

My thrifty parents, ever watchful of the familial bottom line, tended to keep the downstairs hallway light off. That meant a long walk down a dark hallway. Past rooms with dark doorways. Up dark wood-panelled stairs, at last reaching an oasis of light on the landing. Normally, such a journey would be intimidating but manageable for a typical five-year-old. But on this occasion, it wasn't. I'd just watched Disney's *Beauty and the Beast* for the first time.

Now, the dark hallway contained monsters. The dark doorways were gaping maws waiting to devour. At the bottom of the dark stairs, the glimpse of distant light at their summit was a powerful incentive to run even faster. And run I did, developing the anaerobic capacity that would stand me in such good stead in future years, both on the sports field and when facing down dangerous dogs.

My five-year-old self was not aware of it at the time, but as I ran in terror through that old, dark house, I was running back in time. Way back. Back to the beginning.

Our species emerged from across different parts of Africa between 200,000 and 400,000 years ago.[3] By 177,000 years ago, *Homo sapiens* had migrated beyond the continent.[4] As we travelled across a brave new world, we carried with us the imprint of our experiences in Africa. Experiences that included large carnivores.

For our earliest human ancestors, predatory animals, like lions and leopards, represented a number of different things. They were threats to defend against, much like that hulking XL Bully. They were competitors to strive with. And, on occasion, they were food.

In these African landscapes we developed the psychological tools necessary to respond to these animals and survive in these scenarios. We fought; we froze; we fled; we fawned. Our wits were sharpened. Our instincts were honed. And as a social species, whose sometime-cooperative nature complemented our opposable thumbs, this process took place together. Our shared understanding and experience of the world around us shaped us, and then shaped that world.

Back in Ballybay, as I faced the beast that waited in the darkness, these ancient inclinations were very much alive. As I looked out into that corridor of fear, my mind prepared my body for the choice ahead: heart racing, palms sweating, pupils dilating. But this time I did not fight. I froze, and then I fled.

I wasn't just running from the beast in the film, but also from the wolves: that pack of slavering savages whose snarling visage blanched my normally rosy cheeks. And I wasn't just running from both of these predators because they were scary. I was also running from them because they were bad. In the story of *Beauty and the Beast*, and now in my story, they were the villains.

Telling stories is what makes us human. Telling stories helps us make sense of the world, overlaying random chance with comforting categories of right and wrong, of us and them. It was only natural that in the earliest expressions of human culture, where large carnivores held such extraordinary powers of life and death over us, we ascribed to them extraordinary feelings. Feelings like awe and wonder, especially as oral traditions coalesced into spiritual beliefs. But more often than not, these feelings also included fear, anger and threat.

As our stories were canonised around campfires, we added moral dimensions to predators. Wonder became worship. Fear became judgement. Species became symbol. Our inner and outer landscapes had collided.

Ages passed. Civilisations emerged and crumbled. Empires waxed and waned. And our earliest African instincts about large carnivores, distilled through human history, were transmuted to us as stories in oral and written form. We called them fairy tales.

This process of prehistorical oral cultures being transmitted into written history took place mostly in the medieval and early modern era, primarily in Europe. Folk tales from across the world feature charismatic animals, such as 'The Nodding Tiger' from China or 'The King of Crocodiles' from India. But the systematic collection and publication of these tales appears to have been a mostly Western phenomenon, providing us with a distinctly European slant on our modern catalogue of these stories.

Classicism and Christianity begot Europe, even in its increasingly secular form. And Europe, for better or for worse, begot much of the world we know today. Therefore, European stories about predators, whether Little Red Riding Hood's Big Bad Wolf or Goldilocks' Three Bears, have arguably had an outsize impact on how we perceive these animals. These are age-old tales that reflect age-old fears.

They also reflect the age-old vilification of large carnivores, especially wolves. Just like my Viking ancestors were portrayed by my Irish ancestors as wearing horned helmets – not because Viking helmets were horned but because the narrative of the early Christian scribes, who first portrayed these invaders, identified them as demonic, and gave them horns to match. In the same way, predatory animals, despite being amoral, were often typified as immoral. Categories comfort us. And as with all history, they tend to be written into it by the victor. Whether we were threatened by marauding Vikings or by marauding wolves, the fear and loss was easier to bear if you believed you were in the right. And it was easier to believe you were in the right when it was written down on vellum.

Yet deeply embedded within these fairy tales is something more than our instincts and our judgements. Deeply embedded within them is trauma. Intergenerational trauma is the transmission of trauma to the descendants of someone who experiences it.[5] Over the last decade, for instance, our three children have all developed a wariness of strange dogs, not because they've been physically attacked by one but because we've passed our wariness on to them. We haven't sat them down and formally briefed them on how to react; they have just absorbed it from our reactions. Though Joshua has an enhanced wariness because, even if he seemed unaffected at the time in Longstanton, he still carries those memories within him. And after encountering the XL Bully, we did ask him to avoid that part of town when he was walking home from school.

Did our most distant ancestors experience trauma at the hands of lions, leopards and hyenas 200,000 years ago on the African savannah? Undoubtedly. Has this trauma been transmitted to us, and may this have some epigenetic outworking in how we relate to animals like these today? We can probably only hypothesise about the former and, as to the latter, experts disagree about the extent to which intergenerational trauma 'switches' genes 'on' or 'off' in the offspring of traumatised parents.[6] Psychologists disagree about the extent to which fears and phobias are shaped by nature, nurture or a combination of both.[7]

However, aside from our assortment of fairy tales, which pool the collective psychology of thousands of generations, there is one important way in which our experience of competing with carnivores is brought into the present. Its practice is only 10,000 years old but it holds an outsize importance in how we relate to large predators, past, present and future. It is farming.

After hunter-gathering, agriculture came next as a means by which we scratched a living from harsh and unforgiving ecosystems. First in the great river plains of the world as we cultivated, and then in their hinterland as we herded, we competed with wildlife, from parasites to pachyderms. Trauma manifested: from losing a crop to locusts and not being able to feed your family, through to losing your family as they protected the herd from predators. What makes this sort of trauma memory that much more relevant than that experienced by our more distant hunter-gatherer ancestors is that it is that much more recent.

We are all but a few steps away from the land. For the vast majority of residents of Britain and Ireland, it is only 250 or so years since their descendants swapped fields for factories. For many, maybe even for a majority and especially for those outside the old industrial heartlands, it is much more recent.

The same caution we must exercise in inferring intergenerational trauma about large carnivores also applies here. Can we prove that, say, an urban Mancunian whose great-great-great-great-grandfather lost a young bullock to a wolf in the mountains of Carlow is wary of lynx being reintroduced to Galloway because of the trauma of that experience passed down to her through the generations? Again, we can only speculate. But again, the social-psychology legacy of experiences like these live on in our stories, our cultures and our history. Just think of those naughty wolves in *Beauty and the Beast*, chasing Belle on her horse, that had my five-year-old self so terrified.

In the next chapter, we'll take a much closer look at the history of this coexistence with carnivores across these islands. For now, we'll consider how these instincts, thought processes and stories from our past – whether shared round a campfire or via 5G – interact with the latest research on how we think about things, especially in our relationship with the natural world.

The nature of human nature

I am not a psychologist. That will be obvious to the psychologists reading this chapter. But I have done some research on how people perceive snow leopards and their conservation in Nepal.[8] I've benefited hugely from receiving counselling over the years. And I'm married to a psychotherapist, who applies psychology with her clients every week. A lot of my interest in and thinking on the subject I've absorbed vicariously from Paula.

All of this has convinced me that psychology matters. Immensely. We are what we think. And the world has become what we collectively think. Our inner landscapes shape our outer landscapes; the implicit is made explicit.

This holds true for our engagement with the natural world, as we've already charted in evolutionary and historical terms. But what of the present? And what of the future? In particular, what of the various overlapping, and sometimes conflicting, social processes that bring us into direct contact with nature? In the context of this book, these are farming, conserving, rewilding and reintroducing.

All of these are social processes. All of them. Without exception. All of them involve people doing things to, with and for nature based on vision, values and value. Period.

All of these processes are on a continuum of control, or the extent to which humans have ordered and domesticated the landscape, its processes and its species. Yet even amidst the sea of glasshouses in western Holland, one of the most intensively farmed places on Earth and one that we'll pass through soon, there are wild geese eating the grass on the dykes. And even amidst the wilderness of Yellowstone National Park, that rewilding utopia, which we'll also visit later, there are roads and tourists and signs of vanished – and excluded – cultures. Rewilding and farming are merely at different points on the same spectrum of human interaction with Planet Earth.

This book will challenge the comforting categories of wild and domestic, conserved and farmed, intrinsic and utilitarian, that we often rely on to make sense of the world and our place within it. Whether you're a tweed-clad, tea-drinking crusty farmer or an oat-milk-latte-sipping, sandal-wearing bearded bunny-hugger. Or whether, like myself, you're a little bit of both.

But keep hold of this idea of continuum. In the next two chapters we'll investigate more thoroughly these shifting and flexible notions of 'wild' and 'rewild'. In Chapter 12, we'll examine in detail the wide range of interest groups connected to the reintroduction debate in Britain and Ireland and the politics of it all. For now, though, the focus is on the spectrum of self. Just as we can have a breadth of perspectives on any given issue within our society, we can also have a similar breadth of perspectives within our psyche. I'd like to introduce you, briefly, to psychic multiplicity.

That our brain has hemispheres to it, and that these hemispheres can shape different aspects of our behaviour and personality, is not a new concept.[9] Nor is the understanding that our personality contains multiple parts, often negatively associated with dissociative identity disorders like schizophrenia. What is more recent, at least in Western thought, is a theory that multipart personalities are universal, and 'an evidence-based therapeutic model that de-pathologises them'.[10] It's called Internal Family Systems (IFS).

I confess an interest in IFS. My wife Paula is a certified IFS therapist. I've benefited from it personally, finding it more effective than other forms of counselling that I've received. Others think so too: the model has gained significant popularity in recent years, not only as a form of therapy, but also as a type of coaching and even as a form of dialogue and mediation within organisations.[11]

IFS works on the premise that we all have multiple sub-personalities. If you've seen the *Inside Out* movies, think of the various internal characters of the main protagonist, Riley. Different aspects of these sub-personalities, or 'parts' in IFS-speak, collaborate in multiple ways to produce our varying emotions and behaviours. All of these parts rotate around a compassionate core, or 'Self', the essence of our being.[12]

This model helps to explain how we can think differently about the same issue at the same time, even holding seemingly opposed or contradictory thoughts simultaneously. For instance, a part of me enjoys biting my nails and another part realises that it's a sign of stress. Another part criticises me for not having the self-control to stop; another part keeps putting it off. IFS can also help us understand the different behaviours that these parts exhibit, and their origins. My nail biting, for example, may be the result of an early childhood stressor when I found the habit distracted me from feeling the stress. Thirdly, applied in relation to others, this theory can help us appreciate why people think, speak and act in certain ways. Someone else who finds my nail biting disgusting or irritating may be more understanding of this habit of mine if they appreciate that it probably developed as a coping mechanism for stress.

So what does this have to do with sharing landscapes with large carnivores?

First, let's apply the IFS model to some of the themes we've already covered in this chapter.

With dogs, a part of me loves them. Another part is very wary of strange ones, especially XL Bullies, and knows they have killed eight people in the UK between 2021 and 2023.[13] Still another part judges those who keep dogs like these. And another part understands that they may do so to seek protection or validation.

With *Beauty and the Beast*, a part of me loves that film. Another part still remembers the fear it induced as a five-year-old navigating those dark stairs. My 'Self' feels compassion for this part – perhaps if the hallway light had been kept on, I wouldn't have felt so afraid. Other parts feel silly and amused by it all: when the large carnivore conservationist in the room admits they had a childhood fear of *Beauty and the Beast*, everyone laughs.

All of these parts are welcome. All of these parts are valid.

Now let's consider my interest in large carnivores, and their potential reintroduction, from an IFS perspective, setting aside the obvious suggestion that I may still be working out my childhood trauma from *Beauty and the Beast*. Maybe that's why I'm really writing this book. But I do love wolves and lynx and bears. They and their ilk enchant me, especially lynx and the 40 other species of the cat family. They always have. But I also know that they can be difficult to live with, posing real or perceived threats to humans and their activities. Nature, after all, is red in tooth and claw. I fully understand why our ancestors got rid of these animals. I also fully understand why many people want to see at least some of these species return. A part of me feels the same.

Yet I also love cattle and sheep and goats. I love farming as a way of life and a way of being. It's in my blood. Its roots run deep. I know how hard it is to run a small farm on marginal ground, and cope with the cascading demands of market, state and society. I'm deeply sceptical that the British or Irish governments have the capacity or competence to manage something like this. I worry that this issue will be deeply divisive, a feeling compounded by my bitter dislike of sectarianism in all its forms, further informed by the Northern Irish parts of me. I fully understand why many people do not want to see any of these species return to our shores. A part of me feels the same.

I am, though, on the balance of it, neutral on this topic. Yet here there's even a smug part, that feels that sitting on the fence will save me from censure. And there's another part, still, that knows that sitting on a fence, whether barbed, electric or otherwise, can be very uncomfortable. The farming part of me speaks from experience in this area, literally and figuratively. I still bear the scars. Not to mention the entrepreneurial part that has spotted a clear gap in the market to be the calm voice of reason in the midst of a furore.

And that's just a summary of my multiple perspectives on a single topic, albeit one that I spend a lot of time thinking about. Isn't the human mind as marvellous as it is multiplicitous?

Finally, we can apply what we've learned about the IFS model to our collective perspectives on these animals and this topic, perspectives informed by evolution, history and culture. We can appreciate that, as powerful animals, they evoke in us powerful, and often polarised, emotions. We romanticise and we demonise. We eulogise and we curse. We worship and we judge.

No wonder that we often disagree over large carnivores and their nature. No wonder we clash over how best to manage them when they do return, or over whether they should return at all. But if there's one thing I've learned from applying the IFS model to myself, and from thinking about it in relation to carnivore coexistence via 56 interviews and visits across Britain, Ireland, Western Europe and North America, it is that multiple perspectives on thorny issues are real and valid. I should know. I hold most of them myself.

Doing so has also affirmed my belief in the importance of linking conservation and psychology. Apart from some notable exceptions,[14] the two traditions have not had a great deal of overlap, at least in the academic literature.[15] Yet if conservation is the social process that I and many others argue that it is, and rewilding and reintroducing along with it, then understanding the psychological elements of the process isn't optional. It's essential.

This is doubly so with charismatic species like big predators. An increasing number of carnivore coexistence models factor in social aspects, though less so psychological ones.[16] Some even recognise the significance of our collective perceptions of these animals.[17] But in part due to conservation's historical bias towards natural science, there's still a lot of room for refinement.

On top of this, rewilding as a concept also invokes strong feelings in us.[18] Freedom and fear mix with chaos and control. Add the emotions that predators produce in us, and we have a combustible mix on our hands. As we've seen, competing, and even conflicting, feelings about the ins and outs of large carnivore reintroductions, both within and between individuals – myself included – lie at the heart of the issue.

That's why I believe that psychology is one of five key factors shaping this conversation across our islands, and elsewhere. Ecology is but one of them, though it is important and we will explore it in detail in Chapter 11. Psychology and ecology aside, the others are politics, philosophy and economics. Call it PPE+PE for short. It is these five factors that will drive the many facets of the process. It is these five factors that we will keep returning to throughout our journey and this book. And it is these five factors that can help to defuse the debate.

Psychology matters. It made me write this book. It made you read it.

Psychology hones our instincts and shapes our stories. It primes us to respond to real dangers that threaten our lives and to imaginary ones that threaten our happiness. It is personal and it is shared.

Psychology is complex and varied and multifarious.

The psychology of our inner landscapes shapes our outer landscapes, and vice versa. Is there room in these landscapes for wolves and lynx and bears to roam again? To explore that, we need to travel back in time to understand how and why our distant ancestors exterminated these species from these islands. But we also need to understand how and why some of our recent forebears and contemporaries began to think about bringing them, and other extinct species, back. In history, we find the evidence of the implicit made explicit in our coexistence with carnivores.

CHAPTER 3

Where the Wild Things Aren't

The Dutch bison towered over me. It dwarfed even my tall frame, and I knew that one could easily outrun me. My high school 400-metre record? It could eat it for breakfast. And then, being a ruminant, eat it all over again. Not only that, but this female had a calf at foot. I'd been around enough domestic cows with young to know that these were the ones to keep an eye on. Two eyes, if you could spare them.

I recalled my previous close encounter with bison. I'd been working as an animal keeper on the three hoofstock sections of a wildlife park in the Greater Vancouver area. It was the summer of 2008. I was 20, enjoying the second of my long university summers, and feeling like all my Christmases had come at once. I was fulfilling a lifelong dream of being a zookeeper.

As I drove my truck past the free-roaming scimitar-horned oryx and a pair of eastern bongos, I passed fenced pens with a solitary male takin and an addax buck. I was grateful that it wasn't my shift on their section. On the days that I had to clean out their indoor housing, I had to run the gauntlet to it down a narrow walkway. On one side the demented, hand-reared addax thrust his corkscrew antelope horns at me through the chain-link fence. On the other side, at the same time and not to be outdone, the takin – think cattle-meets-chamois or gnu-meets-goat – battered the wire with considerable force, shoulders heaving.

I drove on through the park. I passed the large enclosures with the pair of cantankerous Cape buffalo and then the greater one-horned rhino. Firm fences also kept us well apart. Except for when I had to let them out of their sheds on the mornings that I worked their section. After a quick check inside to locate them, I'd duck though the fence and open the connecting door between their sheds and their outdoor spaces. As I did so, I fervently hoped that they hadn't moved to stand behind the door as I opened it. Wisely, my fight response was nowhere to be seen. Here, with these animals, flight was the only game in town.

But on this morning, on this section, there was one large herbivore that I couldn't avoid standing next to. Without a fence to be seen. On the way to one of the tapir enclosures, I had to pass through a large section of the park with a herd of free-roaming wood bison, close cousins of the more familiar plains bison. Associating trucks and their inhabitants with food, the herd of bison stood inside the gate and looked on as I got out of my vehicle. A four-ton audience. And there were only six of them.

The gate opened towards them, so there was no avoiding what came next: I pushed it slowly forward while gently encouraging these substantial bovines to shift themselves out of the way. They didn't listen to a word I said. Or move a muscle. I retreated to the truck and did the same with the horn, while slowly driving forwards. They begrudgingly backed up. Just a bit. I drove through, parked the truck and walked back to close the gate, while they enquired where their breakfast was. Feet separated us.

'Sorry guys – I only have tapir food with me. You'll have to wait a bit longer for whoever's feeding you today.'

But the bison from my Canadian flashback were captive, accustomed to having people at close range. The European bison, or wisent, that I stood before now was not. It was from a place in the Netherlands where wild bison roamed.

By now, you're probably thinking that I have terrible luck. Attacked by an XL Bully in Chapter 2 and face to face with mother bison and calf in Chapter 3? Hospital clearly beckons by the end of Chapter 4. The grave shortly after, no doubt. You're probably thinking that you should politely decline any invitation to go on a walk with me, such are the scrapes that I get myself into. Especially so because, if you're smart, you're probably thinking that I could outrun you. After all, when faced with running away from a dangerous animal, you don't have to be the fastest to survive. Just the second slowest. But if there are only two of us, and one of them is me, the calculus changes. Then again, you might hope that my protective instinct kicks in and I fight to protect you.

In this case, however, with the giant wisent looming over me, I wasn't worried. For one simple reason: it was stuffed.

I was in the visitors' centre of National Park Zuid-Kennemerland. Located due west of Haarlem, and only a short train ride from central Amsterdam, this is one of the most densely populated parts of one of the most densely populated countries in the world. Yet here be bison, and semi-wild cattle and horses, grazing among the dunes. Here also be tourists – over a million of them each year[1] – and dual carriageways and beachside resorts. Surrounded by urban sprawl on three sides, and the North Sea on the fourth, the park challenges our preconceptions of what is wild, and where wildlife might live. Over there and far away? Or among us and our many activities?

I was here to explore this idea and this place with my research assistant, Jack Crone, on placement from the Leadership for Sustainable Development master's programme at Queen's University Belfast. Jack was an affable former student of politics and philosophy, who had turned his attention to conservation and sustainability; he and I were cut from the same cloth.

As we were baked salmon-pink by the searing June sun, we walked through the wild. Cars zoomed past. Bicycles too. We passed under a living bridge, a special wildlife passageway over the road, one of several in the park. As we reached the bustling seaside, the banks of white wind turbines pirouetted offshore. To the north, we could see the industrial infrastructure of Ijmuiden, one of Amsterdam's ports. To the south, the silhouettes of The Hague's skyline, where I'd been the day before to talk to the LTO, the Dutch Farmers' Union, about wolves.

We turned in this direction, heading past the inner core of the park where the bison lived. As it was off limits to the general public during and after calving season, we were disappointed that we wouldn't be able to see these magnificent animals up close. If it had just been me, I would probably have chanced it. And probably ended up in hospital by the end of this chapter, never mind the next. But with Jack with me, I just wasn't sure that either my employer's or Queen's public liability insurance would cover death or injury from goring or trampling by bison. Worse, my source of free placement students would dry up. We decided to stick to the straight and narrow.

Almost back at the park HQ, we passed a small lake. Local sun-worshippers in various states of undress paid homage to the fire-god in the sky. A few hundred metres away bison roamed.

'I wonder what might happen if the bison escaped from their fenced area and turned up here for a wallow?' I joked to Jack.

We walked on and marvelled at the way nature and human nature overlapped in this sand-dune savannah, in a way that made us question what wild was.

The purpose of this chapter is to make us do just that, by journeying back in time. We're going to do so to consider two important themes. Firstly, just like in National Park Zuid-Kennemerland, we'll explore this idea of wild. If reintroducing is sometimes a component of, as well as a means to achieve, rewilding, then we need to understand what the wild in rewild means. Or, more accurately, what we perceive the wild in rewild to mean. And not only as a concept but, also, as a *place*. We also need to think about how large carnivores fit into all of this, literally and symbolically. Secondly, we'll take a look at extinction, analysing this process in relation to the loss of wolves, lynx and bears from Britain and Ireland hundreds and thousands of years ago.

History has often been absent from conservation. Given its ontological orientation towards the natural sciences, many of those who practise the discipline have equated history with evolution alone.[2] Not only has this relegated much of the last ten thousand years to a footnote, but it has also, at times, avoided awkward questions about the historical role conservation has played in acts of injustice, especially but not exclusively in the Global South.[3] There has been much progress in these areas in recent decades. Yet social scientists are still a minority in the sector, let alone historians.

But if conservation, rewilding and reintroducing are the social processes I argued they were in Chapter 2, then history is everything. History gives context, and challenges and explains the processes of creating stories and science alike. Far from detracting from my role in and experience of conservation, my study of history has underpinned it.

At the same time, social science and the humanities, when applied to conservation, can sometimes disintegrate into so much analysis that it leads to near-total paralysis. I like to get stuff done. There's almost nothing like it. With conservation, getting stuff done doesn't mean we need less natural science. But it will be done better with social science and history as its equal partners.

The past is a fitting place to continue this journey together. The past sets the scene for the future of sharing landscapes with apex predators. History and psychology join forces to explain.

Unearthing the past

Our four-year-old daughter waved at us. We waved back. She was flailing a long stick in the air. Or at least I thought that's what it was. As we got closer I could make out slight protuberances at either end. What had looked like a brown colour from afar was giving way to a sullied ivory. Dust from it fell into the dappled light that spilled through the yew tree behind Bethany. Then I realised what the item really was. It wasn't a stick. It was a femur. A human one.

We were in the old Agivey graveyard in the townland of Aghadowey, Co. Derry/Londonderry, where some of my Irish ancestors have lived since the nineteenth century. A family of badgers had moved in. Being scrupulously tidy sorts, they'd decided to do a little spring cleaning. Out with the old and in with the new. Out with these unsightly bones that littered the substrate. In with a shiny new sett, under the roots of the yew trees.

In her very short career as an excavating archaeologist, Bethany had uncovered more exciting artefacts than I ever had in mine. My own brief stint took place in the summer of 2008 on the Swedish island of Gotland. I was 20, still enjoying the second of my long university summers, and again feeling like all my Christmases had come at once. I was fulfilling a lifelong dream of being on an archaeological dig.

It was quite an experience. For starters, there was the multi-national student team: three Belarussians, two Samoans, two Ukrainians, an English girl and me. Our shared hostel was a melting pot of cultures, cuisines and languages. Serious differences in precedent had to be reconciled. How much salt in the cooking? Were seatbelts mandatory or optional in our hire cars?

Every day for five weeks, along with 20 Swedish undergraduates and the dig supervisors from University College Gotland, we turned up in Västergarn, a half-hour journey south from the island capital, Visby. In the blazing sun, we dug. We brushed. We sifted. We photographed. We sketched. We surveyed. We sweated. We burned. Indiana Jones it was not. My brief flirtation with the idea of becoming an archaeologist was history.

With the benefit of hindsight, I'd do things very differently now. Fence the field. Throw in a family of badgers. Then throw in a gang of small children. Between them, they'd have twice the excavating done in half the time, and enjoy it to boot.

But one day, as I worked in that Gotlandic trench I had been assigned to, I found something. Up to now, it had been mostly animal bones and large stones, probably the remains of a long-vanished barn. As I sieved yet another bucket of sediment, something metallic glinted in the Baltic sun. It was small and round and dusted with soil. As I carefully lifted it, and blew some of the earth away, I could see it for what it really was. It was a coin.

It had probably fallen from someone's pocket and been buried amidst the straw and dung of the barn floor. It had probably lain there for at least 400 years. Undisturbed. And although it wasn't quite the femur of a distant Scandinavian ancestor that I'd hoped for, this artefact was now intimately bound to me. And I to it. Forever.

The hand of history had reached out through the ages and touched mine.

It was a profound moment. I felt deeply moved. I still feel deeply moved.

History is not the past but information about that past. Or so I had learned in History 101 at Queen's University Belfast, two academic years before. Whether by way of primary sources[4] or via archaeological artefacts,[5] we seek evidence to understand not just *what* happened, but *why* it happened.

Yet inevitably, because history is also a social process, just like rewilding and farming, it is always constructed by people with preconceptions and prejudices. People like you and me.

The past is always viewed through the lens of the present. Just like in that Aghadowey graveyard, we can confuse bones for sticks. But just like in that Västergarn field, the hand of history still reaches out through the ages and touches ours. Sometimes it tells us as much about ourselves today as it does about our antecedents yesterday. History becomes not just information about the past, but the creation of imagined pasts.

For history is often contested. That explains why I've written Derry/Londonderry earlier in this chapter, rather than choosing one of the place names. It's a contested term in Ulster, like much of the province's chronology and etymology. It was contested in the past. It is contested in the present. And it will be contested in the future.

It is the same with the wild.

We evolved from nature. We saw that in the previous chapter. To that nature we added stories and created human nature. And of all the stories, it is history that best charts this process of societies continually adapting, and adapting to, the world around us.

But as we also observed in Chapter 2, our perceptions of our outer landscapes are shaped by our inner ones. When it comes to wild landscapes in particular, we have viewed the past through the lens of the present. The prism of two related ideas has distorted our view.

Firstly, and especially in Europe, we became obsessed with closed-canopy forest as the ideal ecosystem of the past and, in much of the rewilding movement, the future. Wild became woodland. Saving the planet and planting trees became synonyms. Complex wood-pasture mosaics, however, which form a habitat continuum between closed forest and open grassland, are arguably a truer depiction of much of the European landscape for much of human history and prehistory.[6]

Secondly, and especially in the Global South, the wild became somewhere where people were not. Halcyonic, human-free zones. Wild landscapes became synonymous with places like the Amazon, the Serengeti and Yellowstone. People were conspicuously absent. Definitely not like National Park Zuid-Kennemerland. Yet again, in each of these examples, there were inconvenient patterns of historical human habitation and use.[7] The absence of *Homo sapiens* from them was a relatively recent, and usually human-induced, phenomenon.

The history of the wild, and of conservation's place within it, was written by the victors. Those who gazetted and managed these places. Or this history was not written at all. Much of the last 40,000 years – of people and nature coexisting in various ways across most of the world's land surface area outside Antarctica – was often relegated to a footnote. The idea of the wild as a place of uninhabited primeval forest, maybe with the odd mountain range thrown in, or, at a push, some savannah, seeped into our psyche. Story became science.

Why is our impression of the wild so at odds with the facts? Or are these just different versions of the same story? History, again, holds the answers.

Romanticism was an intellectual movement of the late eighteenth to mid nineteenth century. Itself a riposte to the reductionist thinking of the Enlightenment, the movement idealised the past and the natural world, as well as our place within them.[8] It prized emotion over function. Here, too, came Jean-Jacques Rousseau's idea of the 'noble savage' – simplistic and problematic tropes about indigenous peoples, their way of life and their relationship with nature.[9] American transcendentalists like Ralph Waldo Emerson and Henry David Thoreau, reacting to the urbanisation and industrialisation of their native northeastern USA, progressed the Romantic idea in North America.[10] In 1872, Yellowstone National Park, the world's first, came into existence.

Romanticism strongly shaped our idea of the wild. Its historic hand has reached down to us through the ages. Yet it wasn't – and isn't – all bad. It counter-balanced an Enlightenment tendency to reduce nature, animals and even people to utilitarian cogs in a machine. It counter-balanced the Industrial Revolution's urge to subjugate natural and human capital to produced and financial capital.

This past was neither utopian nor dystopian; it just was. A shifting mix of science and stories, facts and feelings, evidence and emotion. Just like the present. And the future.

With the IFS lens we encountered in Chapter 2, we can appreciate that we may even hold these multiple perspectives within us, never mind between us. We can also appreciate that differing life situations may have forged different combinations of these feelings within and between us. Some of us may be more utilitarian. Others more romantic. All of these personal permutations are real. All are valid.

Akin to our history of the wild, our history of coexisting with carnivores exhibits this same spectrum of perception. We imprinted on them the complex and sometimes contradictory continuum of beliefs that we saw previously: pragmatic, utilitarian, scientific, rational, metaphysical, spiritual, symbolic.

Similarly, human-induced extinctions of large predators – and herbivores – occurred long before our articulation of the wild as an idea. They occurred at the hands of prehistorical, indigenous cultures. They occurred at the hands of modern, industrialised cultures. So if the wild as an idea isn't directly correlated with the loss of large carnivores in the past, why consider it in such detail? Because, in the context of potential reintroductions of these animals to our islands, it is directly correlated with how we view large carnivores in the present. Almost more than anything else, wolves, lynx and bears symbolise wild.

This particular idea has a firm grip on us.

Finally, our enduring notion of the wild was not just an idea but a *place*. Because it was spatial, it meant that there were places that were wild and places that weren't. Places where wolves and lynx and bears should be, and places where they shouldn't. Large carnivores in ancient, mountainous closed-canopy forest with no people? Tick. The same animals in sand-dune savannahs alongside livestock and people? Definitely not.

We applied our comforting categories to the geography of carnivore coexistence. Not only the us and them, and the good and bad, that psychology gave us. But now also the here and there that history gave us.

In the next section of this chapter, we'll assess this spatial separation of wild and domestic, people and predator, as it drove, in part, the extirpation of these species from Britain and Ireland.

Extinction is forever. Or is it?

Extinction is nothing new. It is a component of the cycle of creative destruction that is evolution. But our species has significantly elevated the natural, or background, rate of extinction through our actions.[11]

This appears to have been true across most of human history and prehistory. The Pleistocene defaunation of the great herbivore herds of Europe.[12] The industrialised slaughter in early modern and modern times as Europe, and those nations and cultures it begot, carried this out on a global and industrial scale with whales, seals and bison, to name but a few.[13] And, of course, the quickening pace of extinction, as the industrial model was systematically globalised after World War II, that has cascaded down through the trophic levels of ecosystems to threaten even insect populations near their base.[14]

This worldwide pattern has held true in Britain and Ireland. The same arc of natural history has bent towards extinction here, accentuated by humans and our nature. From various large deer species in the Pleistocene and Holocene through to great auks in the nineteenth century and on to farmland bird populations in our own day.[15]

Predatory species were not immune from this trend either, locally or globally. Gone were cave lions, cave hyenas and sabretooth cats.[16] Gone were the thylacine – or Tasmanian wolf – from Tasmania,[17] Haast's eagle from New Zealand,[18] and the Bali tiger.[19] And gone were wolves, lynx and bears from Britain and Ireland. A point worth making about the last six of these species is that they all became extinct on islands. The insular nature of these spaces makes populations more likely to become extinct on them, according to something called the theory of island biogeography.[20] It also makes these extinctions permanent, in ecological terms. Only human actions – such as reintroductions – can then restore these species once lost. The rest of our story unpacks the complexities of this process in the two islands I call home.

Wolves, lynx and bears – the stars of our story – are not extinct globally, however. That makes their disappearance from these islands extirpations rather than extinctions. Nevertheless, as with the loss of any species from a landscape, especially keystone ones, there are multiple consequences. We'll look at the inverse of these effects – in other words, the main arguments for bringing them back – in the final part of the book: ecological, economic, political and philosophical. For now, we'll look at each of the three large carnivore species that survived in Britain and Ireland until relatively recent historical times, beginning with bears.

The Eurasian brown bear (*Ursus arctos arctos*), of Goldilocks fame, is a big unit, its bulk further enhanced by a coat of dense fur. More omnivorous than carnivorous, it is ready and willing to snack on most things, from berries to beef and everything in between. I've never worked with bears in captivity, and until this journey, had only snatched a fleeting glimpse of a wild black bear from a mountain road in Canada. But based on the robust construction of bear bins alone – designed to prevent said bears from eating human rubbish – I have immense respect for their capabilities and resourcefulness. Plus that discomfiting ability to stand tall on two legs, and either look us in the eye or even look down on us. They put us in our place.

Our ancestors thought so too. Their physical power and humanoid capacity to walk upright made them useful for entertainment. According to the poet Martial, bears were taken from the forests of Scotland to satisfy the grisly needs of the Imperial Circus in Rome.[21] More robust physical evidence, of bear bones with butchery marks, has pushed back the evidence for human presence in Ireland to 12,500 years ago. Not content to lay claim to 22 US presidents,[22] we Irish have even laid claim to the entire polar bear species, via descent from a separate, extinct species of Irish brown bear.[23] In Britain, bears appear to have lingered longer. They were probably always rare, but there is debate about whether the species finally vanished in the late Neolithic/ Bronze Age – think Stonehenge – or the early medieval period – think Saxons and Vikings.[24] Given the limited evidence, it is hard to ascribe their departure to a single source. But, directly through hunting, or indirectly through competition for space and food, our ancestors probably had a hand in it.

The Eurasian lynx (*Lynx lynx*) – appearing in no fairy tales that I was read as a child – is a mid-sized feline, about the size of a Springer Spaniel. But there's nothing diminutive about its abilities. Like all cats, it is a power athlete. A stealth fighter. Fluid motion. Grace.

As I worked with two pairs of lynx at a Hertfordshire cat collection in the early summer of 2007, I was struck by their intelligence. A brooding sentience behind those baleful, yellow eyes. Compared to all those large herbivores in Canada, I never once felt physically threatened, even when cleaning out their enclosures with them in it. But I did feel looked down upon. Intellectually, at least.

Did our ancestors feel the same? It's impossible to say because the evidence for lynx overlapping with people across Britain and Ireland is fairly limited compared to bears, and especially to wolves. In Ireland, the sole piece of evidence is from a femur found in Kilgreany cave, Co. Waterford, and dated to just under 9,000 years ago.[25] Archaeologists urge caution in assessing evidence from this site, due to the extent of disturbance. In other words, we can't really base the widespread presence of lynx in Ireland on this single piece of bone. But given the cryptic and elusive nature of the species, we cannot rule it out. After all, absence of evidence is not evidence of absence. If that were the case, you could assume a population crash in the Irish Iron Age, compared to the Bronze Age, such is the relative paucity of some types of evidence.[26] In fact and in part, it may be more a result of the damp Irish climate and what that means for the survival of iron remains. Overall, however, the solitary Kilgreany femur does leave an emphatic question mark hanging over lynx in Ireland.

In Britain, there is more physical proof for lynx presence and persistence. Lynx bones from two sites in northern England have been dated to 1,842 and 1,550 years ago, respectively.[27] The authors of this study suggest that a seventh-century Cumbrian text alludes to the species. Yet others point out that no lynx-related Celtic or Old English names live on.[28] More recent work has suggested the presence of lynx in early modern Scottish text which, if accurate, would bring forward the presence of the species in Britain by about a 1,000 years.[29] As with bears, lynx were probably never common. But unlike bears, their size and natural shyness makes it likely that they could have persisted longer in, say, the more remote parts of the island. Either way, the same combination of factors – hunting, plus competition with humans for prey and habitat – probably did for lynx as they did for bears. They vanished from our lands.

The Eurasian grey wolf (*Canis lupus lupus*), of Little Red Riding Hood infamy, lies somewhere between the lynx and bear in terms of size. In dog terms, think a large German Shepherd, Japanese Akita or Alaskan Malamute. Yet the species has an outsize hold on our imagination, our culture and our history. Working at an animal sanctuary in north Antrim later in the summer of 2007, home to, among others, confiscated pit bulls, unwanted reptiles, injured native wildlife and an old tiger called Sonia, I was handed a lead one day.

'Go walk the wolf.'

That was a misnomer. Rather, the wolf walked me. Round and round the old safari park site. Psychotic pit bulls bayed for our blood. Sonia chuffed a greeting. An injured buzzard keened in alarm. It was a surreal experience.

There is an air of the surreal surrounding the history of wolves in Britain and Ireland. In England, the last wolf was reported to have been killed in 1496, but the record of this dates from much later, in 1872.[30] In Scotland, the last official record is from Perthshire in 1680. Throughout Britain, however, as with the other large carnivore species, wolves are likely to have lived on in the more remote and less inhabited parts of the island. The species was definitely perceived as a threat to royal deer hunting, and probably to monastic wool interests.[31]

In Ireland, roughly a third the size of Britain, the species lived on until, officially, 1786.[32] Here, in Co. Carlow, the last Irish wolf was shot for rustling sheep. Other evidence points to their relative abundance in Irish geography and history, including a slew of place names across the island.[33] The most common archaeological feature in the Irish landscape, raths or ringforts, often more loose corrals or stockades than fortified positions,[34] may have been designed to deter not only human raiders of cattle, but lupine ones as well.

The surreal element of wolves as larger-than-life creatures, symbolic and even metaphysical, is also apparent in Irish history.[35] There were rumours of werewolves. In the era of Oliver Cromwell, certainly not a popular figure in Ireland, they acquired particular political and economic dimensions.[36] Although bounties on wolves had been issued prior to the mid seventeenth century, this systematic and professional Cromwellian persecution of the species, designed in part to accelerate the settlement of seized land, sped up the species' decline.[37] As in Britain, and as with lynx and bears, shrinking habitat and prey populations, coupled with hunting, doomed the wolf. So too did a confluence of political and economic factors, especially the threat the species posed to valuable resources.

With lynx and bears, it's not clear that our fear of either species – the us and them, the good and bad of our psychology – played a significant part in their extirpation. With wolves, it probably played more of a role, as we will also encounter in the present, later in our story. But a struggle for space, and the ability to control who and what inhabited it, certainly did, especially in Ireland. Nor is it clear that perceptions of the wild shaped this contest between people and predators. That is a later lens through which we view these parts of the past. Yet the spatial element of the wild – the here and there of our history and our geography – excluded these species.

Where the wild things are became where the wild things aren't.

Now, the hand of history is reaching out through the ages. But this time, it is our own hand. It is our hand reaching back to firmly grip some version of this wild past and these wild places, real or perceived, and yank them firmly into the present. Along with these wild species that symbolise them most.

For it turns out extinction may be forever, but extirpation is not.

For some, rewilding and reintroducing is turning the clock back, while for others it represents moving forward. For some it is utopia, and for others, dystopia. In the next chapter, we explore the origins of these powerful trends that are shaping our islands and our world.

CHAPTER 4

Comeback Kids

My muscles burned. My lungs too. Every step was agony. Every step upward required concentrated power of will. I was in Nepal's Sagarmatha – or Everest – National Park (SNP) on the trail of snow leopards and a controversial reintroduction proposal. But first, I had a mountain to climb.

It was early February 2014. Even before a successful scoping trip in autumn 2013, I had applied well in advance for the various research permits required. One, for my other field site, Annapurna Conservation Area, had come through in good time. The other one, for Sagarmatha, remained as elusive as the big cat whose conservation I was in Nepal to study.

My PhD research team and I had assembled in Kathmandu a few days before. There was no SNP permit in sight. After much discussion and verbal approval from the relevant ministry, we decided that the rest of the team would travel on to the Everest region to begin fieldwork, while I remained in the city to track down the paperwork. I spent a few days camped outside and in various offices in the main government compound. Eventually, the relevant official showed up. I emerged triumphant waving the all-important piece of paper.

I headed straight to the airport, anxious to catch up with the rest of the team. After the bone-jarring, death-defying flight to Lukla, gateway to SNP, I emerged intact, if a little paler and queasier. Flying into one of the world's most dangerous airports has that effect. It tends to focus the mind on matters of life, death and immortality. I'd even sent a brief farewell message to Paula and the kids as I boarded. Just in case it was my last.

After a brief shot of caffeine to steady my nerves in the local Irish pub, I set off. Not only was I trying to fit two days of walking into one to make up for lost time in the field, but I was also carrying a rucksack of 24 kilos, a personal record. And while the distance from Lukla to Namche Bazaar – where the rest of the team were already hard at work – was only 15 kilometres, the continuous descent and ascent into and out of steep river valleys made it feel like many more.

As I retraced the path that the rest of the team had taken only a few days before, the route got progressively more difficult. The last three hours, on a steep ascent from the main river valley, were a real killer. A combination of altitude and payload took its toll. Every step upward required agonising effort.

I should be entertaining you here with a riveting description of one of the world's most iconic landscapes. I should be waxing lyrical about the view from the famous bridge over the Dudh Khosi river, festooned with prayer flags, that appears in every Everest film. I should even be linking all of this physical splendour to metaphysical musings on the nature of life, death and immortality, as Himalayan odysseys seem to elicit in those who attempt them.

Nothing could have been further from my mind.

Instead, my world constricted. My focus narrowed.

Breathe. Step. Pause.

Breathe. Step. Pause.

Over and over.

On and on.

So I can only entertain you with a riveting description of the way the dust danced beneath my tired feet in the fading light as it poured through the pines. I can wax lyrical about how the sweat flowed from my pores, down my face, stinging my eyes, no matter how often I wiped them. The only metaphysical musings were four-letter ones about mountains, PhDs, bureaucrats and snow leopards. Not to mention the mystical blue sheep and its proposed, but contested, (re-)introduction to the area.

Dusk fell. Eight hours after leaving Lukla at 2,800 m, I staggered into Namche Bazaar at 3,400 m.

I've walked among and climbed mountains on four continents. This remains the hardest day's hiking that I have ever done.

Reintroducing species can be like this journey in many respects.

Difficult. Unglamorous. Resource-intensive. Contested.

Our primary goal through our fieldwork was to assess coexistence between local people, snow leopards and their conservation across the Everest and Annapurna regions, which we did by surveying 700 people and interviewing 70. As part of this work in the former place, we were assessing local attitudes to the proposed movement of blue sheep, or bharal, to the area. Taxonomically speaking, and despite its name, the blue sheep is a wild ungulate species somewhere between a sheep and a goat.

The premise was simple. In SNP, with wild herbivore species like Himalayan tahr and musk deer relatively thin on the ground, snow leopards were taking some livestock. Elsewhere, blue sheep, including in Annapurna, are one of the snow leopard's favourite meals. Plans were being mooted to move a population of this magnificent creature from other parts of Nepal to the Everest region, in the hope that snow leopard predation on livestock would decrease. A study to check if the habitat was suitable for bharal had already proved to be positive.[1]

There were just two minor problems. One, nobody had asked the locals what they thought of these plans. Two, there was conflicting evidence about whether the species was, in fact, native to the area after all.

On the first point, we aimed to remedy that. Our questionnaire results from 260 households in the Everest area showed moderate support for the idea. However, our more nuanced findings from interviews with 26 local leaders were revealing.[2]

The tourism potential was 'good as people might come to see the animal', noted a local microcredit cooperative officer, while a teacher theorised that 'as the snow leopard's preferred prey, it may help to reduce livestock losses if its population is growing well'.

Less positive remarks focused particularly on the potential crop damage from the species. 'If it also destroys the crops in the way the [Himalayan] thar do then it's not good to translocate it... In that case people won't accept blue sheep here,' said a youth leader.

That those most affected are often the least consulted is an unacceptably common refrain about wildlife reintroductions. This is partly due to conservation's historic bias towards natural science, which we've already discussed in previous chapters. But it's also partly due to the perception that 'pristine' wild places ought not to have people living in them in the first place. Or so goes the caricature of this outdated view.

Being a fan of underdogs myself, as well as a conservation social scientist, consulting those most affected was the reason behind talking to locals in the Everest area about this blue sheep translocation idea. It's also part of the motivation behind this book, and its accompanying research report,[3] on large carnivore reintroductions to Britain and Ireland. In my opinion, consulting those most affected is the place to start, not end.

As we consider the history of rewilding in this chapter, focusing especially on the history of reintroductions across the world, Europe, Britain and Ireland, this is a theme that we will encounter repeatedly. It's an indicator that rewilding and reintroducing, like history itself, are often contested processes. And emotive ones.

Back in Nepal, the second and most significant aspect of the debate was whether bharal belonged in the Everest area at all. Some sources have recorded historical sightings of blue sheep in the area. One recent trekking map, for instance, noted blue sheep sightings further along the trail to Everest Base Camp. Others question this, pointing out that even the species' presence in the neighbouring Gaurishankar Conservation Area was historical and not current.

Some locals concurred with this assessment.[4] 'We should focus on conserving the species here and the blue sheep in its own habitat. Introducing it will hamper the conservation of other species here,' observed a teacher. Similarly, a conservation leader remarked that instead of 'artificial conservation with blue sheep we would be better to think of ways to increase the number of [Himalayan] tahr'.

This debate hints at the complexity of reintroductions, and their often uneasy relationship with history, which we will consider throughout this chapter. If blue sheep did exist in SNP previously, then their return from, say, Gaurishankar or from neighbouring Tibet is a re-colonisation. The animals bring themselves back, in response to changing ecological conditions or expanding populations. If they never occurred in the Everest region, then this process would be a colonisation for the same reasons. Species – including our own – spreading across the globe in search of space and resources is a key process of life on Earth.

However, if blue sheep did occur previously in the national park, and people brought them back, it would be a reintroduction. Although this may seem like semantics, the fact that people would be the ones bringing the species back changes everything. It changes it from a natural to a social process. And all social processes – as history shows us – are contested.

If blue sheep were never native to the area, then this process becomes even more contested. It would no longer be a reintroduction, but a novel translocation, something that official guidelines frown upon,[5] and for good reason.

In the case of blue sheep, the process of introducing the species to SNP appears to have stalled. But this example raises all sorts of questions that we will encounter in this chapter on rewilding history, especially as it relates to potentially reintroducing wolves, lynx and bears to Britain and Ireland.

As we've already seen in our discussion of what 'wild' means, 'rewild' is even more loaded with meaning. And with emotion. This is especially the case if this process involves people returning species previously extinct in an area. Doubly so if the species in question have sharp teeth and pointy claws.

Who is affected and who is consulted? Who pays for it? How does the historical evidence change the case for reintroduction between, say, the 9,000-year-old lynx femur in the Co. Waterford cave, on one hand, through to the relatively recent wolf record of 1786 from Co. Carlow, on the other? Are lynx reintroductions to Ireland, in particular, a reintroduction or a novel translocation? Who gets to decide?

As we consider in this chapter the history of rewilding as an idea, and its manifestation through reintroductions globally, regionally and locally, we will begin to consider the answers to some of these questions. We'll also unpack them in greater detail throughout the rest of the book.

Like my uphill trek to Namche Bazaar, our journey through this landscape promises to be challenging but unforgettable. Also like my trek, it comes with a lot of baggage.

(Go) fly a kite

My muscles burned. My lungs too. The radio antenna I had strapped to my rucksack was an awkward shape, though not heavy. This time, though, the mountains were considerably smaller – mere speedbumps. And the altitude was negligible – just above sea level.

But mountains are mountains. They humble the human spirit. And the human body.

You're probably thinking that I'm a bit of a glutton for punishment, running around mountains with heavy things strapped to my back. In fact, as I wrote the first section of this book, I was training for a return visit to Nepal. Part of that involved carrying a 25 kg pack around a lovely but challenging six-mile route through Glenariff Forest Park in Co. Antrim, just over the hill from where our journey began in Chapter 1. Every weekend, usually with my son in tow.

But there is something very liberating about it. Especially when you're carrying all that you need to survive on your back – food, water, shelter. It is primal. It is freedom.

On these training hikes, however, it's been four sleeping bags and two weight plates that I've been carrying. Nevertheless, the effect is the same. A physical challenge. A mental workout. Freedom.

In fact, I've been enjoying it so much that I may keep it up even after my Nepal trip, as often as I can. I thoroughly recommend it.

Mountains have always called to me. Though I was born in Ireland, I was made in the mountains of Malawi. These were the landscapes that I spent some of my younger and all of my high school years in. These are the landscapes that speak to me most.

As it so happens, mountains are usually amongst the wilder landscapes on the planet. Places where agriculture and other forms of human activity are less intensive. Places where wildlife clings on.

For these very same reasons, mountainous and upland areas are often where rewilding and reintroducing are most likely to take place.

That's why I was lugging a radio antenna around the mountains of Mourne in Co. Down. As I systematically volunteered my way through as many different types of wildlife as I could during my university holidays, one group remained elusive, despite my best efforts: birds of prey.

But a former manager at a Belfast nature reserve that I spent a year volunteering at once a week, while working towards an NVQ in conservation, became the manager of a project to reintroduce red kites to Northern Ireland. I jumped at the chance to lend a hand during Easter 2009.

Between 2008 and 2010, 80 birds were reintroduced to the Mourne landscape, the first time red kites had been resident in Ulster for over 200 years.[6] As I toured the area on foot and by vehicle, taking readings from the birds' radio tags via the antenna and then noting individual IDs via binocular and spotting scope, I gained a sense of the process.

I watched the kites hang on the thermals, sentinels in the sky. Their beautiful forked tails contrasted with the buzzards' blockier ones. The deep reds and greys of their plumage shone in the spring sunshine.

There was something marvellous about this species' return. A second chance. Redemption, even.

Where the wild things aren't had become, once more, where the wild things are.

There is a universal element to this. The theme appears again and again in films, books, even the greatest sporting events. Everybody loves a comeback kid.

But my youthful enthusiasm for the process was being tempered. It began to dawn on me that not everyone was happy with the return of the kites.

Some birds had been shot.[7] Others poisoned. Still others had simply disappeared.

Local farmers had had concerns that red kites might predate on lambs. Whether valid or perceived, these concerns were still real to those who held them and may have been behind some of the illegal killings.

I learned from my brief experience of this project that a reintroduction wasn't just a straightforward story of a wrong being righted. It wasn't just a process of the tame becoming a little more wild again. And it definitely wasn't a binary situation of the 'good' conservationists versus the 'bad' farmers.

It was infinitely more complex than I realised. And infinitely more contested.

This nuanced view of reintroductions that I developed among the reintroduced red kites soaring above the Mourne mountains has remained with me ever since. These themes – complex and contested – also appear throughout the history of wildlife reintroduction efforts.

Rewilding wasn't a term that had appeared on my radar at this stage, despite immersing myself in the world of conservation as often as I could. Nor was it a term that I encountered during my short stint with the red kite project. So how do the terms rewilding and reintroducing relate to each other? And do they both relate to the idea – and place – of the wild that we encountered in the previous chapter?

Rewilding as a term came to prominence in North America in the 1990s. A famous paper boiled it down to cores, corridors and carnivores.[8] The catchy phrase, and its explanatory alliterative trio, stuck. Large areas of relative wilderness connected to each other by corridors of less wild habitat would allow wide-ranging populations of large carnivores to move freely across landscapes, exerting transformative ecological effects down through the levels of the food chain. We'll discuss more of these effects, or trophic cascades as they're called, later in the book.

So far so good. Several decades on and we can still see wild spaces and wild species as important components of rewilding. Arguably, it is the presence, movement through and return of wild species, especially big predators, to the spaces outside and between these wilder cores that pose problems for people. Firstly, this is because there are more human activities here that conflict with large carnivore presence, especially livestock farming. But secondly, it is also because if we perceive the wild as a distinct place where certain species may belong, such as a national park, then their presence in and especially return outside of these places challenges our notion of what and where wild is.

Rewilding as a term has also morphed from this early definition to a more plastic, or catch-all, phrase that can be deployed to mean a wide variety of things by a wide variety of people.[9] From reintroducing elephants and rhinos to Britain and Ireland, right through to planting wildflowers in the middle of the village roundabout, and everything in between. This plasticity also allows its detractors to define and perceive it in multiple ways. From a subversive attempt to undermine the hallowed neatness of Little England's sacred lawns, through to radical, fruitcake-crazy, even dangerous, efforts to remove people, and especially farmers, from the landscape. Take your pick.

As with the strong feelings about large carnivores that our evolution and culture provoke, it's no wonder rewilding as an idea also evokes intense emotions. Nature means as many things to us as our human nature allows us to imagine.

And as we saw from our consideration of the IFS model of psychic multiplicity in Chapter 2, we're pretty good at having multiple perspectives on a single issue within, never mind between, ourselves.

But rewilding doesn't just force us to consider *what* and *where* wild is. It also forces us to consider *when*. In other words, it adds a temporal dimension to the conceptual and spatial elements of wild that we've already considered. It does this by adding two letters to the word 'wild'. These two letters, 're', mean 'anew' or 'again'.

The familiar hand of history is now reaching not only into the current, but forward in time as well. Suddenly, past, present and future are all connected on this continuum of wildness. For many, this is a lot to take on board. It is also at the heart of why rewilding is a complex and contested term.

Many of us take very little interest in the past, though it takes a considerable interest in each of us. I was even warned by one prospective publisher not to spend too much time on the history of this topic in case readers got bored. As a historian, I tried very hard not to be offended. As it so happens, apart from odd people like me who are interested in the past from a very young age, I find that people tend to become more interested in the subject as they grow older. Maybe it's all those metaphysical musings on life, death and immortality.

But, regardless, we all – without exception – take a healthy interest in the present. And in the future.

Hope springs eternal, as Alexander Pope famously quipped. Hope is future-focused. The future matters a great deal to us; it's the stuff of hopes, dreams and aspirations, as well as fears, worries and trepidations.

Rewilding brings the wild past, or at least our perceptions of it, smack bang into our present and our future. For some, it butts heads with their vision for that future, and the values that underpin this. For others, it is a return to a primordial past, where people are put back in their place, or even removed entirely, at least from certain spaces. For still others, it is a journey towards a dynamic and exciting re-evaluation of our relationship with ourselves and the world around us. Even within the rewilding movement itself, these competing visions and definitions exist. It is a complex and contested continuum on multiple axes: control versus chaos; old versus new; wild versus domestic; self-willed versus human-dominated; dystopia versus utopia.

Add the potential return of animals like wolves, lynx and bears as both instrumental and symbolic parts of the rewilding process, and it's easy to see why this topic generates a lot of heat. As we've seen from the original North American use of the term in the 1990s, predators were a clear part of the vision. Whether they deserve the role of ecological saviour anointed by the likes of the 'wolves-change-rivers' story from Yellowstone, and whether these same effects can be transposed to very different European contexts, is something we'll explore in greater depth later on.[10]

As we saw in Chapter 2, we all bring our polarised perspectives on predators to the reintroduction debate. Just as there is a part in many, if not most of us that still fears predators, there is also a part in many, if not most of us that still worships them. We may not be comfortable with those terms. We may not even be consciously familiar with these feelings. But it's probably within each of us, buried deep inside. If you can't buy that, then perhaps you can accept that this paradox of predator perception certainly lives on in our stories, whether by the Brothers Grimm or Netflix. It certainly lives on in the rewilding and reintroducing debate.

Whether they return by themselves through recolonisation or with human assistance via reintroduction, they bring with them this baggage. Or, more accurately, we attach it to them.

Rewilding and reintroducing are loaded terms. But how have these been applied in the real world, with multiple, overlapping and competing visions for landscapes and the species within them, including our own? In the next part of the chapter, we'll expand on our brief history of reintroductions.

Try to stay awake. It's anything but boring.

Second coming

That the idea of rewilding began in North America in the 1990s is not surprising. Thanks to the legal protection afforded by the 1973 Endangered Species Act, the funding that this unlocked and leveraged, plus campaigning by dedicated individuals, rewilding had its most famous and controversial expression here. In 1995, wolves were reintroduced to Yellowstone National Park.

Depending on whom you ask, and ask I did as I travelled through Colorado, Wyoming and Montana, this was either a symbolic act of biological repair that unlocked not only profound ecological benefits, but significant economic ones that continue to this day. Or a gross overreach by the federal government, spurred by the influence of wealthy, urban and out-of-state actors who didn't have to live with the consequences of wolves spreading far beyond the park.

In Chapter 11, we'll not only take a closer look at this example, but also visit Yellowstone's famous Lamar valley, and go on a guided tour with a wolf ecologist. For now, it's fair to say that this example has had an outsize impact on the rewilding debate across the world. If I had a pound for every time someone in Britain or Ireland mentioned Yellowstone when I told them about my book-and-report combo, I would be rolling in it.

Closer to home, a similar combination of causes facilitated a no less significant, but less widely recognised return of large carnivores, including wolves. The 1979 Bern Convention provided legal protection for these species across the European continent. This was followed in 1992 by the EU's Habitats Directive.[11] The subsequent EU LIFE programme provided strategic and systematic funding for biodiversity conservation across the bloc, including many reintroduction efforts.

Three decades on, rising from historic lows in the 1950s to 1970s,[12] the brown bear population across Europe is between 17,000 and 18,000,[13] while for wolves it is around 17,000,[14] and for lynx 8,000 to 9,000.[15] Most of this recovery has been through natural recolonisations, aided by those legal protections, by funding, and by dedicated individuals and organisations, as these animals have returned to parts of their historic range. Sometimes, reintroduction projects have complemented or sped up these efforts, including the return of Eurasian lynx to Switzerland,[16] Iberian lynx in Spain and Portugal,[17] and brown bears in the Pyrenees.[18]

Predators aside, Rewilding Europe has emerged as a leading proponent of nature recovery across the continent. From their headquarters in the eastern Dutch city of Nijmegen, the organisation works directly in 10 focal landscapes, and indirectly through more than 90 initiatives across 28 countries in the broader European Rewilding Network.[19] As we'll see when we visit Oostvaardersplassen Nature Reserve in the next chapter, the closest thing Europe has to Yellowstone in terms of fame or infamy, you tend to find the pioneering Dutch popping up to lead rewilding innovation, as they do in so many other fields.

In Britain, and even more so in Ireland, the rewilding process has been slower and less complete. Large carnivores remain a subject of passionate debate only, mainly due to two bodies of water called the English Channel and the Irish Sea. As we saw with the theory of island biogeography in the previous chapter, that made our island populations of wolves, lynx and bears more vulnerable to extirpation. It also made this extirpation permanent in ecological terms. Only humans can change that.

With its well-established environmental movement, rewilding in Britain began with discussions and trials by some of its lesser-known members, including the Wildland Research Institute and the British Association of Nature Conservationists.[20] But two books – George Monbiot's *Feral* and Isabella Tree's *Wilding* – arguably popularised the idea, the latter deeply grounded in what has become Britain's best-known rewilding project, Knepp. Rewilding Britain mirrors Rewilding Europe's role as a network of projects across the island, while particularly in Scotland, with its lower population densities and concentrated landholdings, organisations like Scotland: The Big Picture and Trees for Life have advanced the vision.

The type of animals most commonly reintroduced to Britain over the last few decades have had pointy claws but not pointy teeth. Birds of prey including red kites, ospreys and sea eagles now fly in our skies again, as recounted by Roy Dennis in *Restoring the Wild*. Over the last decade or so, beaver reintroductions – both licensed and guerrilla – have captured the public's imagination, as Derek Gow's *Bringing Back the Beaver* details. Not everyone has been pleased by these reintroductions, including farmers worried about damage to sheep flocks by sea eagles[21] and landowners concerned over drainage issues relating to beaver ponds.[22] More recently, a small herd of bison have been successfully reintroduced to Kent as part of the Wilder Blean project, with the aim to use the species as a woodland manager.[23]

When it comes to large carnivores, most of the British rewilding books – and I ordered the bulk of the canon in preparation for writing mine – make passing reference to wolves, lynx and bears. Most, however, are very short on detail as to how coexistence with livestock farmers would be managed and governed, other than vague assurances that it works well in Europe so it could work well here too. The Lynx UK Trust, who began proposing trial lynx reintroductions to Thetford and Kielder forests from 2015 onwards, had a licence declined by Natural

England in 2018. The regulator's official report is fairly damning about the lack of consultation with local communities, landowners or farmers.[24] I was designing and teaching a short course on human–carnivore coexistence for conservation master's students at Queen's University Belfast when I first came across their proposals. To me, this lack of consultation seemed arrogant, top-down and technocratic, a textbook example of how not to do it. In fact, I resolved there and then that should the opportunity present itself in the future, I would look into the topic from the bottom up, putting farmers' concerns at the forefront. Eight years on, this book, and the accompanying technical report, is the result.

And it's not only me who had concerns. A leading British rewilder whom I interviewed told me that their approach had set back lynx reintroductions to Britain by at least ten years. Rewilding Britain, themselves generally in favour of lynx reintroductions, expressed their concerns about some of the project's proposed sites, their estimate of potential sheep losses and the level of consultation with landowners.[25] Thankfully, the latest and current initiatives to actively explore lynx reintroductions to Britain – Lynx to Scotland and the Missing Lynx – led by a consortium of rewilding organisations, have a much more humble and down-to-earth approach. We'll look at their and the Lynx UK Trust's contrasting proposals in greater depth in the third section of the book, as together they comprise the most tangible steps taken to date to realise this vision in Britain.

In Ireland, by contrast, discussions about large carnivore reintroductions, never mind plans, are nowhere near as advanced. That said, a number of studies assessing ecological viability have emerged recently, some of them acknowledging that public consent, and especially farmers' concerns, need to be placed front and centre.[26] We'll also consider some of their findings in the final third of the book.

But, like Britain, Ireland has witnessed some raptor reintroductions, including golden eagles to Co. Donegal, sea eagles to Co. Kerry and, of course, the red kites we encountered in Co. Down. Just prior to the time of writing, two sea eagles who turned up in Co. Antrim from Scotland were illegally poisoned,[27] a reminder that not everyone is happy when these species return, whether they do so by themselves or with human assistance.

As with species reintroductions, rewilding as a process and movement is not as well established in Ireland. A number of organisations and sites, including the Wild Nephin project in Co. Mayo, Co. Meath's Dunsany Nature Reserve and Co. Clare's Hometree, are all pursuing rewilding innovation in their own way – though a formal rewilding network, as found in Britain and Europe, has not yet sprung into existence. This is partly a reflection of the historic political weakness of the Irish environmental movement, relative to its British counterpart. But the times they are a-changing. As the Republic of Ireland catches up to and surpasses the UK on a wide range of indicators, including GDP and public budget surpluses, and as a greater environmental consciousness develops in the Irish public writ large, the rewilding movement here is likely to develop faster. There's also the small matter of access to EU funding for such projects, which has been such a significant facilitator of rewilding across Europe.

Call it rewilding, nature recovery or ecological restoration: the process holds a powerful attraction for many, and provokes an equally powerful rejection from others. As we've seen from our encounters with blue sheep, red kites and Yellowstone, this spectrum of colour is matched only by the spectrum of opinions. To a large extent, these hinge on where the wild as an idea belongs in practice, and on what point on the spectrum of human-ordered landscapes we are willing to accept a certain amount of 'disorder' brought by non-human species.

Rewilding is also a process, a particular vision of the future and of the landscape. This often conflicts with other visions, especially if it involves the return of powerful predators that evoke equally powerful emotions. Over and above any actual impacts – ecological, economic or otherwise – these animals have a far greater symbolic impact. This is due, in large part, to their significance as symbols of the wild as an idea, of the wild as a place, and of rewilding as a process. They are lightning rods for passionate arguments for and against.

Speaking of passion, it's also worth concluding this chapter by pointing out that rewilding and reintroducing as social processes are often driven by passionate people. Consider the pantheon of rewilding saints briefly mentioned in this chapter. Their charisma and determination gets things done. Or the interesting people who ran the animal collections that I spent my university summers working in. Some of them I still admire, some of them I thought a bit odd, and some I was downright scared of. But their charisma and determination still got things done.

Charismatic animals, likes wolves, lynx and bears, tend to attract charismatic people to their cause. This isn't a bad thing. I happen to be one of them. But it's worth acknowledging as a factor in our discussions. They can be as polarising as the species in question, adding additional complexity to the social – and therefore relational – process that is human–wildlife coexistence.

These relationships – between people *and* wildlife as well as between people *over* wildlife – can be managed and governed through coexistence strategies, including a suite of management tools and governance processes. But what does coexistence look like in practice and where should it occur? And can rewilding take place without large carnivores, or are they essential for it to be fully realised? Chapter 5 investigates.

As we journey forward together, we bring more baggage with us than on a Himalayan expedition: evolutionary, cultural, historical. It's the same with all social processes, reintroductions included. It's part of what it means to be human.

Onwards and upwards.

CHAPTER 5

Missing Links

Our train sped across the savannah. Limitless expanse reached to the horizon and, surely, beyond it. In the distance, we could make out herds of herbivores moving through this famous landscape. I'd always wanted to visit the Serengeti. Except I wasn't in Tanzania. I was in the Netherlands. And this was the Dutch Serengeti. This was Oostvaardersplassen.

Hopping across the country from our visit to Zuid-Kennemerland National Park the previous day, my research assistant Jack and I had stayed the night in the fairly new town of Almere. Built on a reclaimed polder from the 1970s onwards, the municipality still had a frontier feel to it. As we ate in McDonald's that evening, we watched a drug deal go down just outside.

After a night in a very dodgy hotel, we boarded the train for Lelystad. The town was at the north-east end of the park, and one of the main gateways to it. Though our taxi driver had never heard of the place. Maybe that was just down to our butchering of its pronunciation. In fact, I'm still at a loss as to how to say Oostvaardersplassen correctly. My valiant efforts have elicited either polite smiles or noble attempts to rescue the Dutch language from my monolingual massacre. All have failed. Now, I stick to calling it the Dutch Serengeti. Or that rewilding place in the Netherlands with the unpronounceable name.

Like both Almere and Lelystad, Oostvaardersplassen was built on land reclaimed from the ocean. Beyond its horizon was the Markermeer, a 700-square-kilometre lake that had once been part of the North Sea. Like Zuid-Kennemerland, the park was bordered by water on one side, and either urban or agricultural land on the other three. And like the sand-dune savannah of the previous day, this seemed like a strange place for the wild.

Urban skylines intruded on the horizon. Wind turbines pirouetted here too. And the familiar linear infrastructure of the Dutch landscape – power lines, railways, roads, canals – bisected the land.

We were visiting the frontier of rewilding. In the canon of rewilding books that we galloped through in the previous chapter, Yellowstone may often be Exhibit A. But Oostvaardersplassen is frequently Exhibit B.

The place wasn't just the Dutch Serengeti. It was the European Yellowstone. I expected to be tripping over not only multitudes of large herbivores but also swarms of rewilders on a pilgrimage to the process's spiritual home on the continent. I reckoned I'd probably encounter at least three other authors writing books on rewilding. I was disappointed on all counts.

As we walked and talked, Jack and I frequently encountered evidence of large herbivore presence. The sign and smell of horses were ubiquitous. It reminded me of family safaris in Africa during my teenage years. Especially in Malawi's then depopulated parks, we spent a fair amount of time staring at bushes. Willing dancing shadows to animate into animal life.

Yet that is the reality of most wildlife-watching. Nature is not TikTok, where you can scroll onwards if your interest hasn't been sufficiently piqued in three seconds flat. Nature has a different rhythm, one that we have to slow down to appreciate.

Then Jack spotted something. 'Horses!' he cried. Naturally, the horses bolted. We bolted after them.

I caught a glimpse of a tail and a fetlock disappearing into the vegetation. A sea of reeds to which the wind whispered. We even climbed a small mound hoping for an aerial view. But the horses were gone. Or, more accurately, they were gone from our field of vision. Probably less than 50 metres away from us, they had slipped out of our line of sight and into our imagination.

Oostvaardersplassen is very much a product of imagination. Firstly, the pioneering Dutch imagined that they could reclaim land from the sea. So they did. Secondly, the pioneering Dutch imagined that some of this land could become a large-scale experiment in rewilding, instead of another industrial zone. So it did.

We become the stories we tell. And the dreams we imagine.

This was a story of people stepping back from the ecological micromanagement that characterises so many European landscapes and nature reserves. I recalled hours and days spent in nature reserves in and around Belfast, whacking and cutting vegetation. The blisters notwithstanding, it was usually tremendous fun, a welcome change from being immersed in the medieval era – and the library – four days a week.

On one memorable occasion, a gate to one tiny nature reserve we were thrashing bracken in was left open. Suddenly, a herd of Friesian heifers appeared amongst us. We quickly chased them back out of the gate and securely fastened it. But far from wrecking the place, I wondered if that brief trampling-and-browsing session by 30 large bovines may have been exactly what that beautiful little ecological museum needed.

It brought to mind a sign at a community allotment that I used to run: 'I fought the weeds. And the weeds won.'

My experiences fighting vegetation on behalf of conservation were making me question the efficiency and effectiveness of using human labour, even the free university student kind, to this end. Inevitably, the vegetation always won.

What if, given the absence of wild herbivores in some of these places, we threw in 30 Friesian heifers instead? Or 30 goats? Or even 30 pigs? Not only would we accomplish twice the amount of vegetation management in half the time for a fraction of the cost, but they would even add value, as livestock have always done, by transforming plant material that we cannot eat into animal material that we can.

My interest in conservation grazing was born. It brought together my twin loves of farming and conservation. Later, I found this philosophy reciprocated in the practice and process of rewilding. Or at least some examples of it.

Oostvaardersplassen imagined this process on a colossal scale.[1] Thousands of Heck cattle, Konik ponies and red deer all managing, browsing, grazing, trampling, digesting and excreting. Sculpting new terrestrial ecosystems where once the North Sea had lapped the Dutch shore.

It was imagination run wild. To some, it was a dream manifested. But to others, it became a nightmare.

In 2018, after an especially harsh winter, over half of the 5,000-odd animals on the 5,000-odd-hectare reserve were starving.[2] Most were culled by the Dutch state organisation responsible for the park. Many people were outraged. Some even resorted to throwing bales of hay over the perimeter fences as supplementary feed.

There are two inconvenient truths that this story of Oostvaardersplassen brings to the forefront of our story.

Firstly, the fact that death is a natural part of life. Weather and disease have significant effects on wildlife populations, as do predators, both natural and human.[3] Just like a part of me was disappointed that the grazers of Oostvaardersplassen weren't throwing themselves at me left, right and centre, another part of our idealised perception of the natural world forgets that for many sentient organisms, life is nasty, brutish and short. Once again, nature, after all, is red in tooth and claw. This is all part of that evolutionary process of dynamic equilibrium, of creative destruction that underpins the natural world. How quickly we forget. How quickly we airbrush it with sentiment. How quickly science becomes stories.

Secondly, as much as Oostvaardersplassen was like the Serengeti or Yellowstone, symbolically and ecologically, there was one major difference: there were no large carnivores. If there had been, the initial small numbers of ponies, cattle and deer would probably never have reached such unsustainable numbers. But in their absence, humans would have had to fulfil – or at least try to fulfil – that role by harvesting animals from the reserve. That would have made it, at least to my eyes, part farm, part nature reserve. This may have been philosophically unacceptable to those rewilding pioneers who envisioned the project, and its idea of 'self-willed nature' with as little human intervention as possible. Certainly, until their recent return to the Netherlands, reintroducing wolves to the park to perform this role would also have been unacceptable to most. But so too was letting nature take its course in the winter of 2017/18.

Only difficult choices were on offer here. But these are choices that, in a crowded world where the increasing number of opinions collide over a fixed amount of land, have to be made nonetheless. And they are choices that have to be made, to one extent or another, about every landscape, whether it is an industrial zone or a rewilding reserve.

Yet the relative – albeit brutal – logic and simplicity of natural cycles frequently become misunderstood and contested when viewed through these lenses of competing human visions and values. To what extent should humans be involved in the management of rewilding ecosystems, either through farming domestic large herbivores or hunting wild ones? In the absence of large carnivores from these places, including Britain and Ireland, to what extent can or should we function in the role of apex predator? Do we really need wild ones back? And how do our competing human belief systems, which refract all of this through them, coexist, as well as existing alongside non-human species?

In this chapter, we'll consider these questions.

Back in Oostvaardersplassen, a cold wind began to blow. We could taste the North Sea in the air. June suddenly felt like March. Despite the controversies, I left full of admiration for the rewilding pioneers who had conceived of the vision for this landscape, and had possessed the determination to bring it to life. In the shadow of the human race, here was the wild made flesh: as an idea, as a place, as a process. Here was imagination run wild.

The wild hunt

Something moved. As I surveyed the broken hillside through my spotting scope, I caught the briefest flicker of movement at the very edge of my field of vision. I began tracking back to bring it clearly into focus. Someone else had also spotted it. Next to me, the hunter unlimbered his bow. It was the last day of archery hunting season in Wyoming. I was on a hunt in the hills above Lander.

Or to be more precise: I was on the 'glassing' phase of a hunt, a sort-of reconnaissance mission. With rifle season due to begin the next day, the group of hunters I was with were scouting the lay of the land. They were scoping out the presence of elk in this area, thereby increasing the likelihood of a successful hunt in the coming days.

The composite bow and crossbow with us were just on the off-chance that we'd encounter some game. The bear spray cans on our hips were a reminder that we weren't the only predators stalking this prey. The whole experience was new to me.

I'd been hoping to accompany Jaden Bales on an elk hunt to get a feel for the process and the people. In fact, when Jaden had suggested in a WhatsApp message that he might have me come glass with him if I was finished with my Friday afternoon meeting in Lander at a reasonable hour, I thought this was some euphemism for getting drunk. Dutch courage for the first-time hunter. Despite being Irish, and in the middle of developing my own conservation gin brand called Malawi Mountain® Gin, I wasn't a big drinker. I wondered whether I'd be able to keep up with these wild hunters.

But between our WhatsApp conversation earlier in the week and that Friday, Jaden had got his bull elk. Perhaps it was just as well. I was a terrible shot anyway. And as I discovered when 'glassing' on that Wyoming hillside, I didn't have the patience to sit for extended periods of time looking through a scope. As much as I enjoyed the backcountry scenery and camaraderie, I'd rather be moving through it than sitting in it. Preferably with something heavy strapped to my back.

I'd sought out Jaden to help me answer the question of whether humans could function as apex predators through hunting. In other words, if we hunted more in Britain and Ireland, much like in North America, could we bring about the same ecological benefits, namely reducing deer populations, without the complexities of reintroducing large carnivores? Jaden worked for the Wyoming Wildlife Federation, an NGO founded in 1937 to educate the public about conservation. They were enthusiastic advocates for the North American, or sustainable use, model of conservation – aka 'use it or lose it'. Over breakfast, Jaden described how they saw 'hunting and fishing as a vehicle for people's engagement with nature'.

Rewilding without large carnivores requires more human involvement in ecosystem management. If the large herbivores employed in rewilding projects are domestic, then for both legal and ethical reasons, a degree of intervention to control their population may be both necessary and desirable, as we saw in Oostvaardersplassen. Just as wild species and processes in agricultural landscapes challenge, say, farmers' assumptions about where the wild belongs, equally, allowing the farming of meat and other products from livestock in rewilded landscapes may challenge rewilders' assumptions about where the domestic belongs. I personally think it's a good idea to maintain some human food production in most of these places, underscoring our myriad connections with the land, in ways both instrumental and intrinsic.

But if the large herbivores are wild, then hunting rather than farming becomes the means to do this.

Back on the outskirts of Lander, I'd come round to Jaden's to help him butcher his elk. The imposing rack of antlers sat propped against a cottonwood tree. The sun shone through them to illuminate the red rock walls on the other side of the narrow valley. He showed me the picture of him posing proudly with the bull. It was an impressive specimen.

But the trophy hinted at a major problem with my theory about humans hunting like large carnivores do. In short, we're after different animals. Humans, naturally, want the biggest and best. Usually mature males with headgear that looks good over the fireplace and sounds good over a drink.

'Do hunters suffer from the same affliction as fishers?' I joked to Jaden. 'The one that got away that was the ultimate prize?'

He chuckled and nodded. 'If anything, they're worse!'

By contrast, predators tend to focus on weaker individuals within a prey population: the young, the old, the sick.[4] In this way, they help to maintain a healthy balance of the species they depend on. Predators also exert another important influence on prey: they shape their behaviour. Their very presence in a landscape affects where herbivores prefer to graze and browse, in turn affecting plant growth in these areas. It's called the landscape of fear,[5] something we'll consider in more detail in Chapter 11. And it's not something that humans replicate in quite the same way, though the presence of human hunters can certainly alter deer behaviour, for example.[6]

Large carnivores shaping the total populations of deer and the like as well as shaping where those populations are – their behaviour – is at the core of the ecological case for lynx and wolf reintroductions to Britain and Ireland. In our chapter on ecology, as we explore Yellowstone's northern valley, we'll delve into this issue. For now, it's fair to say that for us to have a similar effect on British and Irish deer populations, we would probably need to increase the amount, and change the focus, of hunting.

This raised more questions for me than the number of flies buzzing around my head. As we cut and dressed the saddles of venison in the autumn sunshine, the fly activity increased with the air temperature. Eventually, we called it a day and headed inside to continue our work at the kitchen table. For lunch, Jaden cooked up a storm with the elk heart.

As we ate, Jaden spoke about the legal requirement to harvest meat from deer and elk hunts in Wyoming. Commercial hunting was not allowed, though commercial guiding of hunts was. For those who didn't want or need all the meat they were required to take from their quarry, several schemes to channel surplus meat to food banks had been set up, including one by the Wyoming Wildlife Federation.

I wondered how this would fly with the British and Irish public. Here, hunting has more cultural, even class, baggage. Perhaps it dates all the way back to the establishment of deer parks by the Norman kings, with the harsh enforcement of their hunting privileges. Or perhaps it is more recent perceptions of deer stalking as an elite pursuit on aristocratic estates. Add in the admirable public concern for animal welfare, a proud British legacy, which almost everybody wants considered in principle but almost nobody wants to pay for in practice. Either way, these views arguably colour discussions of controlling deer populations. But with the right incentives, more culling of deer across both islands could simultaneously provide ecological benefits and a source of food.[7]

It wouldn't be quite as effective as wolves and lynx doing it, but it would be less complicated, as we'll spend the rest of this book discussing.

Predators – and their predation – generate diversity and various ecological benefits. Humans – with or without our livestock – struggle to replicate the same benefits in quite the same ways. Yet we can derive and provide many other benefits from enhanced management of landscapes, wherever they may sit on the wild spectrum, such as food, fuel and leisure, to name but a few.

In many ways, it was easier to manage semi-wild places like Oostvaardersplassen, where the large herbivores were technically domestic and boundary fences restrained their ability to range. I realised there were likely to be as many perspectives on the management, farming and hunting of wild and rewilded landscapes as there were on large carnivore reintroductions themselves. And given that wolves, lynx, bears, mountains lions, bobcats, coyotes and wolverines all exist in western Wyoming alongside hunting, it didn't mean that it had to be an either/or scenario. As the Wyoming Wildlife Federation pointed out, large carnivore conservation and big game hunting could, under the right conditions, coexist in the same places.

The short answer to my question of whether humans could function as apex predators was 'not quite'. But did that mean my answer to the parallel question of whether we should reintroduce wolves, lynx and bears to Britain and Ireland was 'yes'? Before I could answer that, I needed to explore how coexistence between large carnivore conservation and other human activities was managed and governed in other parts of the world. And before I could engage with that process fully, I needed to begin with two additional questions: what does coexistence mean and where should it happen?

My experience of the wild hunt with Jaden and his friends had left me with a new-found appreciation for hunting and hunters. I identified with both the social element and the physical challenge, as well as with their love and respect for the natural world. I also had a deeper insight into the challenges large carnivores faced when hunting, namely finding their prey, never mind catching it.

I left as an enormous amber moon rose into the Wyoming sky. The lights of Lander flickered beneath it. I journeyed on.

Sharing time and space

What is this term 'coexistence' that I bandy around so much?

Its Oxford Dictionary definition is: the state of being together in the same place at the same time.

Human–wildlife coexistence, to give it its full title in this context, is a term that has started to compete with, and in some cases supplant, an older phrase: human–wildlife conflict.

The beef some conservationists, and not just those troublesome social scientist ones, had with this phrase was that it wasn't based on ecological reality.[8] 'Conflict' is not a term from nature's lexicon. Secondly, it ignores the complex and competing social, political, cultural and economic dynamics that are often at the root of both threats to and solutions for wildlife.

So human–wildlife coexistence, which is frequently shortened to just coexistence, emerged. It acknowledges, on one hand, human–wildlife impacts in both directions. In other words, people having impacts on wildlife, as we have done from the beginning: hunting, habitat loss, domestication, climate change, etc. And wildlife having impacts on people, as they also have since time immemorial: crop raiding, livestock rustling, home invading, death, etc. And we're talking here not only about the charismatic megafauna – the lynx, wolves and bears of the world – but also the humble termite, mouse, even spider.

On the other hand, and distinct but not unrelated to these impacts, human–wildlife coexistence also acknowledges human–human conflicts. In other words, competition between people, usually in some sort of social grouping, *over* wildlife. Competing stakeholder groups. Contrasting visions and values. Contested versions of the past and the future.

At first glance, this may seem like another semantic, academic debate that could carry on until the cows come home. No doubt it will. But words have agency and power. They are the stuff of stories. And stories? Well, we become them. Stories are the stuff of societies.

In short, human–wildlife coexistence says: if you want to save nature, you've got to understand human nature.

But coexistence is not some utopian, fairy-tale paradise. At its best, it is grounded in the messy realities of a complex and crowded world.

It seeks answers – and questions – outside the narrow confines of the natural sciences that have dominated conservation. Coexistence admits that it does not involve the absence of conflict. Rather, it accepts that conflicts are part and parcel of the process. So too is compromise and, where possible, consensus. Even if the consensus on, say, the topic of this book is that there is very little consensus. That's still a place to start.

Coexistence is journey, not a destination.

And on that journey, like on any journey, especially on this journey, various aids are required to help us on our way.

There are a suite of various technical tools that can be used to optimise coexistence between humans and wildlife – in effect, minimising the negative and maximising the positive. In the light of potentially reintroducing big predators to our islands, these are deterrence, finance, force and enterprise. We'll spend the middle section of the book exploring each of these in detail.

But a narrow focus on these tools is not enough. In fact, relying on these alone, as many human–wildlife coexistence projects often have, is to fall back on the old, tired and technocratic 'conservation as a science experiment' trope.

We need to talk about governance.

Governance is the framework within which we fit coexistence, impacts and conflicts. Governance has space for clashes, compromise and consensus. Governance factors in contrast, competition and contestation.

A focus on governance as the foundation of human–wildlife coexistence pushes forward into the new, vibrant and realistic 'conservation as a social process' endeavour.[9] In other words, it becomes much more than just a series of scientific experiments held together with string. But guess what? There's still plenty of room for science experiments in this big 'conservation as social process' tent. In fact, they're absolutely essential.

We've driven an electric car since 2015. The ethics made sense for us, as did the economics: what we saved in running costs, courtesy of Northern Ireland's then free public charging network, paid for the car over five years. Our humble Nissan Leaf also doubled as the farm vehicle at Jubilee Farm. At one time or another it carried live goats, live turkeys, bales of hay, straw and wood shavings, veg boxes, pork deliveries and even tubs of whey from a cheesemaker. Not to mention three human kids and their baggage on daily school runs. We even took it all the way from Northern Ireland to the Netherlands for the wedding of some friends, despite it having a range of only 65 miles.

On our journey towards coexistence, think of that poor, long-suffering electric car as governance. A vehicle for all sorts of unusual occupants and fellow travellers, some of whom are not happy about being stuffed in the boot. Never mind the frequent stops required on long journeys for charging, caffeine and the loo.

Now think of our various coexistence management tools as components of the car that can be employed to respond to various road conditions. The window-wipers of deterrence. The lithium-ion battery of finance. The brake of force. The accelerator of enterprise.

But there are two things the Jubilee Farmmobile cannot do. It cannot tell us our journey's destination. For that, other factors are required to draw a roadmap: politics, philosophy, economics and ecology. We'll look at each of these in a chapter in the book's final section.

The second thing it cannot do is tell us what roads to take to get to our journey's end. Put another way, where exactly should this human–wildlife coexistence happen? Only in wild places? Or in all places, including where humans are present and perhaps even dominant?

This is a good point in time to briefly consider the land sparing/sharing debate.

More academics arguing to the nth degree about another boring topic, you say? Maybe. But it's not boring. And it's very important.

In a nutshell, the debate hinges on whether we should intensify agricultural production in concentrated areas, mostly the fertile lowlands, thereby freeing up other areas, especially wilder, wetter, drier and higher areas, for nature conservation. Those who favour sparing argue that this creates less, but bigger and more effective, spaces for wildlife.[10] On the other side, land sharing seeks to make space for wildlife in farmed landscapes, through more common but often smaller pockets of habitat.[11]

I used to be firmly in the land sharing camp. Then I set up and ran a small, agro-ecological farm, and came face to face with the pros and cons of that approach. Now, I'm open to the merits of both arguments. In fact, as with most issues in both farming and conservation, there isn't really a one-size-fits-all approach. Sparing versus sharing exists on a spectrum, with the balance varying from landscape to landscape.

A similar debate takes place at the global level. The late E.O. Wilson, he of the *Theory of Island Biogeography* fame, popularised the idea of setting aside half of Planet Earth for nature.[12] But, say its critics, where to put all the people who depend on the half set aside for nature for their livelihoods, especially farmers, fishers and hunters?[13] Again, I find the idea of spectrums, a common theme in this book, rather than dichotomies, helpful here as well.

In reality, the wild – if we take that to mean non-human nature as opposed to only romantic visions of primeval forests – is everywhere.

Under our fingernails. In our gut. In our gardens.

Deep in the ocean. Deep underground. Deep in our psyche.

In cities. In suburbia. In the countryside.

Wild is everywhere. Nowhere is fully wild. Nowhere is fully domestic.

Interplay between these spectrums of the wild and the domestic, the ecological and the agricultural, has defined human history. It also defines the delicate coexistence between people and their activities, on one hand, and large carnivores and those who conserve them, on the other, especially regarding livestock.

But it still doesn't answer the fundamental question: where do large carnivores, these most iconic symbols of the wild, belong? In the next two sections of the book, we'll try to answer this question. In short, perspectives vary wildly.

Because as I travelled across Western Europe and North America, seeking answers to this question, and to how deterrence, finance, force, enterprise and governance could work in practice, a curious case of biological NIMBYism ('not in my backyard') unfolded.

When I interviewed those in the eastern lowlands of Britain and Ireland, they thought these animals might fit in in the western and upland parts of Scotland, Wales and Ireland. In these places, I was told 'definitely not'. Try the forests of Europe.

In the forests of Europe, some Dutch people said wolves belonged in the mountains of Europe. In those mountains, some Swiss reckoned North America was their natural home.

So I went to North America. In Colorado, one rancher recommended Wyoming as the place for wolves. In Wyoming, quite a few people were annoyed that there were wolves outside Yellowstone National Park. To them, that was their natural home. Maybe if I'd visited Alaska or northern Canada, this circle would have been squared. Perhaps here I would have found the backyard where people were entirely happy to have predators present. Then again, maybe not.

For each biological NIMBY, I also met a biological YIMBY ('yes in my backyard'), someone who welcomed big predators back, or their continued presence in the landscape.

And for each of these opposing viewpoints, I met a pragmatist working towards some form of coexistence, setting aside the debate about whether large carnivores should be there and accepting that they were.

In the next section of the book, we'll meet many of these individuals and explore what it means to share spaces with large carnivores – and each other. I'd confirmed that, whether through farming or hunting, whether in places as diverse as Oostvaardersplassen or Wyoming, we couldn't fully replicate the role that apex predators play in ecosystems. There are missing links in our landscapes left by the departure of lynx, wolves and bears. But would the ecological benefits from their reintroductions outweigh the social costs that would presage and accompany their return? In the book's final section we'll consider this more fully.

I'd become convinced that coexistence was as much about sharing landscapes with people as it was about sharing them with nature. In Britain and Ireland, coexistence means being together in the same place at the same time. With wildlife, wherever the landscape is on the continuum of wild. With each other, wherever we are on the continuum of wild. Wherever we sit on the spectrum of opinions about farming or hunting in wilder landscapes; about sharing land or sparing land; about setting aside half of the Earth for nature alone; about the extent to which we can function as apex predators; about the extent to which we need the real ones back.

There is a fixed amount of space on Planet Earth. In the past, as we saw in Chapter 3, this tended to involve our species pushing many other species, including and especially large carnivores, to its margins. Chapter 4 discussed how, in the present, the trends to rewild and reintroduce have attempted to redress this. But, as we saw in this chapter, on a busy and crowded planet, sharing these, often changed and more crowded, spaces with returning wildlife involves navigating a range of perspectives and requirements, with other species and, especially, with our own. Central to of all of this is our inner landscapes. Whether individually or together, how we think and feel inside us shapes the world around us. Chapter 2 considered that if we could hold multiple opinions within us on any given issue, not least apex predators and their potential return, we could surely also hold these between us.

This is the nub of coexistence.

And if we are to share the fixed amount of space on Planet Earth with our own and other species; if we are to share the fixed amount of space in Britain and Ireland with each other and with wildlife; if we are to learn how to live again with wolves, lynx and bears wherever that may be: then coexistence is the missing link that will make or break it all.

PRESENT

CHAPTER 6

The Lynx Will Lie Down with the Lamb

The phone rang. I could see an international prefix on the WhatsApp screen. *Looks like a Swiss number to me,* I thought.

I answered. A woman's voice, in a broad Cumbrian accent, asked, 'Are you Jonny?'

'I am indeed.'

As a I tried to reconcile the Swiss dialling code with the distinct northern twang, the voice continued: 'I'm a farmer in Graubünden, Switzerland, where we're having real problems with wolves.'

This was my first introduction to Sarah Zippert.

'They're not just killing sheep, but cattle now, too. Farmers are really angry. And the thought of someone wanting to bring wolves back to the Cumbrian fells makes my blood boil.'

Once I'd fully reassured Sarah that I wasn't personally planning to reintroduce wolves to Cumbria, but that I was on a fact-finding mission to understand how Swiss and other farmers managed to live alongside large carnivores, much of the tension dissipated. In fact, she'd heard about my planned visit from the local branch of the Swiss Farmers' Union, whom I'd contacted to enquire about visiting some local farmers affected by wolves.

We spoke for half an hour. Or rather, Sarah spoke and I listened as she told me about her beloved Alpine farm, which she managed with her

Swiss husband Kurli, both of whom worked as ski instructors in the winter. And about the rising tensions in the province and country as emotions over wolves and their impacts ran high and boiled over into protests.

By the end of our conversation, I'd been invited to stay with them during my upcoming Swiss trip, their guesthouse vacant during the tourist off-season.

I hung up, delighted to have made such a useful contact. But I was slightly taken aback by the level of emotion that obviously pervaded the issue of wolf presence in Switzerland. What was I going to encounter?

I found out a few weeks later when Sarah picked me up at Chur train station.

A blonde whirlwind of formidable energy and opinions, Sarah spent the next three days introducing me to a wide range of local characters. There was a local biologist and conservationist who'd previously been keen on wolves returning to Switzerland but was now very sceptical. There was the local branch of the Swiss Farmers' Union, based at the provincial livestock market outside Chur, where we stopped so Sarah could pick up a campaign placard that read, in German, 'Regulate wolves. Protect children and livestock'. And, for balance, there was even a visit to the local office of a conservation NGO active in wolf conservation, where the interview, at times, threatened to veer off into a heated debate between Sarah and the conservationist.

But straight from the train station, we first headed six miles down the road to meet Stefan Geismann.

Sarah, Stefan and I were standing in a forest glade on a steep hillside above Untervaz. Dappled sunlight filtered through the trees. The air thrummed with the heat and life of an early May afternoon, as this mountain ecosystem awoke after its long winter slumber. Insects buzzed and floated and crawled. A raven croaked.

Stefan was employed by the local agricultural college to provide technical advice and support services to local farmers in the canton – or province – on livestock protection. He had taken us to see one of his three herds of meat-producing goats in a wooded pasture about ten minutes' drive above Untervaz on a narrow, winding mountain road.

We heard the goats before we saw them. A whistle from Stefan was followed by a clamour of bells as the herd surged up the hill towards us.

The sudden atavistic feeling I had at being surrounded by goats again was strong – so strong and visceral that it was almost palpable. As various bits of my clothing were nibbled and scratched, I was immersed again in the sounds and smells and sights of working with goats.

I remembered the nervous excitement when our new herd first arrived on the farm. I remembered the joy of new kids being born, an everyday miracle that never got old. I remembered the emotional rollercoaster of nursing sick animals, ecstatic when they pulled through and sorrowful when they didn't. I remembered the frustration of repeated escape attempts – a pair of the wee blighters even turning up in someone's garden in the neighbouring village once – and the sheer exhaustion at having to head out yet again to round up wandering stock and patch permeable fences.

I was surprised at how intensely I remembered goats. And I was surprised at how intensely I missed them.

But what surprised me most was the strong protective instinct I felt towards Stefan's goats in this moment. Against wolves, against bears, against lynx: pity the poor carnivore that chose this moment to strike. Hell hath no fury like a farmer protecting their stock. I may have been armed only with sticks, stones and a few choice curses, but any pointy-toothed predator wouldn't have stood a chance.

Suddenly, I was struck by a strange sensation that I'd first encountered when our goats had arrived at Jubilee Farm all those years before. As the setting sun cast lengthening shadows over the marbled blues of Larne lough and the chequered greens of Glynn wood, I watched them slowly walk in single file across the main field in front of the farmhouse that May evening. I felt an upwelling of emotion. And an upwelling of memory. Old, old memories, some of which were not my own.

I felt a deep love for these animals. I felt a proud pastoral duty towards them. And, most curiously, I felt that these feelings, in a way that is hard to articulate in words, were not just mine towards this small herd of goats on that spring day. Beyond that, I felt somehow that this pastoral sense transcended time, becoming, in that moment, a deeper extension of a near-universal human experience.

There was the legacy of multiple generations of my own family who had kept and cared for livestock. There was a legacy of multiple generations of farmers who had tended stock before me on this very land. There was even a sensation of being directly connected to 10,000 years of agricultural heritage, and how livestock had helped us survive, and sometimes thrive, in this world.

This great weight of history bore down on me. But it was not a burden; it was a legacy of life.

It was a feeling I'd never forgotten. I had experienced it on several occasions when working with or even seeing livestock, mostly, but not always, my own. In this moment, with Stefan's Alpine goat herd, I felt it again.

A part of me was experiencing that deep connection we have with livestock and their husbandry. For some of us, that's an ongoing experience. For others, it is but a few generations gone. And for all of us, whether we like it or not, it is part of the legacy of human life on Planet Earth.

An equal part of that legacy has been protecting our livestock – and our livelihoods – from the threats they face. If the livestock losses across the last ten millennia could be calculated, we would probably find, as we do today, that the majority of them come from factors like disease, bad weather, theft and the inherent tendency of domestic animals to do stupid things, like wander off cliffs or fall into rivers. As the old pastoral saying goes: 'Where there's livestock there's deadstock.'

Predators would likely be responsible for a relatively small proportion of these losses. But because of the evolutionary instincts and cultural heritage that we encountered in Chapter 2, these relatively low levels of *actual* risk are refracted through this lens and become much greater levels of *perceived* risk. In other words, intangible factors like fear,[1] compounded by our tendency to have our emotions and values reinforced by our social groupings,[2] are very significant here.

But as vitally important as they are, we cannot just sit around analysing human emotions in this debate. We also need to understand what practical steps can be taken to reduce the risks of farming with carnivores. In this chapter we'll consider five methods: fences, guard animals, shepherds, corrals and collars. We'll blend science and stories to convey the often harsh, and always tricky, realities of reconciling livestock farming and predator conservation.

However, because predation is one of many factors farmers must consider when looking after their animals, we also need to consider deterrence within the broader context of husbandry. Defined by the Oxford Dictionary as 'farming, especially when done carefully and well', husbandry speaks of 10,000 years of human dependence on livestock; of age-old relationships between the farmer and the farmed; and of equally age-old struggles with the multiple threats that disrupt this civilisation-defining partnership. Husbandry is also a defining issue in how well people and large carnivores coexist.

Put simply, farmers losing less livestock to the old enemies of disease, theft, weather and, to use the technical term, 'livestock-being-stupid', are often, but not always, losing less livestock to predators. When conducting my PhD research is Nepal, I found this to be true with snow leopards, who were responsible for 3.4% of livestock losses in the previous year, as reported by local herders.[3] In the Swiss context, about 6% of livestock losses during the summer grazing season are due to large carnivores,[4] though the number of wolf kills have risen substantially since this figure was calculated.[5]

Within this broader landscape, how can farming with carnivores happen?

It's a dog's life

At the end of our second day of visits in and around Chur, Sarah's brother-in-law David came round for dinner to discuss his experience of keeping livestock protection dogs.

We feasted on a hearty Swiss stew of wild venison, shot by Kurli the previous autumn in the woods above the Zipperts' summer pasture. This was solid, home-cooked fare, exactly what I needed after a long day in the mountains.

My glass tinkled as a local wine was poured into it.

'Why did you start keeping guarding dogs?' I asked David.

'Back in 2017, a lone wolf came into our barn and killed three of our milking sheep. Although it was only a single individual passing through the valley, we realised then that we needed to proactively protect our sheep herd – and cheese business – from the possibility of future wolf attacks.'

I enquired if he had previous experience of working with dogs.

'No, and it's been a steep learning curve.'

David went on to recount the ins and outs of working with the dogs. At the time, only two breeds – Pyrenean Mountain Dogs and Abruzzese Shepherd Dogs – were approved by the Swiss government as official livestock protection breeds. With limited availability, getting hold of the animals took some time. Each young dog also had to be trained and then pass a test before it was dispatched to a flock, to ensure it had the necessary protective skills and instincts.

When they arrived, they ate him out of house and home. I remembered the Great Dane we'd had growing up in Malawi. Even as a puppy, she ate more than the Rottweiler and Labrador crosses we also had put together. That said, a monthly payment from the government mostly covered the costs of David's feed and veterinary bills.

What it didn't cover, though, was time. Without the recommended 6,000-square-metre paddock to house the dogs outside the summer grazing season, David spent a considerable amount of time walking them.

Between two and six dogs were stipulated for each flock, to ensure greater protection against a pack of wolves, or in the event of one dog wandering off. David shared how on several occasions, both dogs had gone AWOL.

Sarah chipped in: 'I was driving up to the summer pasture and I noticed David's dogs sitting at the side of the road. I called David and we managed to get them back up the hill to the sheep.'

I asked David if he'd had any problems with aggression from his dogs towards other dogs or tourists, or complaints about barking in the village, all things I'd heard as common problems with livestock guarding dogs.

'Not towards people but they don't like other large dogs, especially huskies or similar breeds that look like wolves. But we're fortunate that we can fence our summer pastures in sections so that walking and mountain biking trails don't pass through them.'

Not every shepherd has this option, leading to conflict. On one hand livestock guarding dogs exercise their protective spirit, drummed into them by both nature and nurture. On the other hand, a broad range of human groups use these same mountain landscapes for leisure, including hikers and bikers, and they also underpin the local economy. It was a microcosm of the coexistence conundrum. How to reconcile multiple overlapping uses of the same piece of land, be that conservation, tourism or farming?

To solve this problem, warning signs were erected on trails and apps were even created to notify tourists of the location of livestock guarding dogs. Yet I'd spoken to one contact in France who, having witnessed first-hand a husky-type dog being severely mauled by a livestock guarding dog, no longer took her own large dog into the high pastures during the summer grazing season. Another Swiss conversation referenced a neighbour's daughter being chased on her moped through the village by an escaped livestock guarding dog.

'But do they do the job they're supposed to and deter wolves?' I asked David.

'We haven't had any wolf attacks since we started using the dogs. At the same time, we haven't had a resident wolf pack in the valley.

As wolves move up from lower altitudes, the real test for the dogs is still to come, maybe this season or next.'

On the way to the train in Langwies the next morning we passed the field where David's dogs were on duty with their sheep. From a distance, it was hard to tell which were the sheep and which were the dogs. Both looked like white bundles of fluff. Maybe that was part of their cunning plan to trick unsuspecting wolves into a false sense of security. Yet the moment we got out of the car, the two dogs stood up, started barking and began moving towards us. I was glad that there was a four-foot electric fence between us.

Despite all the complexities of using them, the dogs are, along with fencing, one of several minimum requirements of the Swiss government's livestock protection – and, crucially, compensation – scheme. Livestock guarding dogs, like their wild wolf cousins, were here to stay in Langwies and much of the rest of the Alpine region.

All of the various deterrence methods to protect livestock from predators work by disrupting the various stages of a predation sequence.[6] This process begins with a predator searching for suitable prey, passes through identifying, approaching, attacking and killing them, and ends with consuming. Fences and corrals – predator-proof enclosures usually used at night – act as barriers to the approach phase. Collars – often of leather or plastic and placed around the neck of livestock – aim to protect this most vulnerable region during the kill phase, especially from cats that, quite literally, go for the jugular. Shepherds and guarding animals, by comparison, are more flexible. By their very presence, their sounds and their actions, they are able to interfere with the approach, attack and consume stages.

All of these methods suffer from a lack of analysis of their effectiveness.[7] Even meta-analyses – studies of studies – struggle to be definitive, although one suggested guarding animals and lethal control, which we'll consider in Chapter 8, were roughly tied in first place.[8] Overall, it appears that the suite of deterrence tools may vary widely in their usefulness, depending on factors like location, species and culture. In short: it's complicated.

But this in no way means we should just give up and abandon all deterrence methods because they are overly complex and under-researched.

On the contrary, each has its own balance of pros and cons. Each also has to be weighed up against the time, money and labour available to implement them.

Let's start with fencing. If you've ever put up a fence, you'll know it can be tricky to get it right, even on flat ground. Jubilee Farm was on the side of a hill, and fencing there was a challenge at times. But that hillside above Larne lough was nothing compared to the one above Untervaz, where Stefan's goat herd was. And, according to Stefan, this was an easy spot to fence compared to some of the pastures his farming clients had to enclose.

Not only that, but Stefan said the wolves were learning to jump over the electric fences. Or squeeze under the gaps that were impossible to avoid as fences passed over rocks, roots and other uneven surfaces. A veritable coexistence arms race was underway, between two highly intelligent species.

Electric fencing was one of a number of deterrence methods required by Swiss law for farmers to be eligible for compensation should they lose livestock. At least one method, from a shortlist of fences, dogs or shepherds, had to be in place. Advisory and support work by extension agents like Stefan supported individual farmers at the local level, while at the national level, centralised research, development and training was coordinated by an organisation called AGRIDEA.[9] The costs of the fencing materials were covered by the Swiss government, but not the costs of installing them. Yet a Swiss study, like many others assessing deterrence methods, was unable to determine the effect of these fences on rates of livestock loss to predators.[10] On the other hand, another larger study of multiple methods across multiple countries found that fencing could be effective.[11]

From Stefan and Sarah's point of view, the fences required a considerable amount of additional time and effort. Time and effort that, when running a farming operation, large or small, part- or full-time, is in short supply. I sympathised with them. When I ran Jubilee Farm, my daily routine – of managing the farm, as well as the social enterprise that owned and operated it – was balanced on a knife-edge. Something as simple as a delay in submitting an application form could mean the difference between finishing the afternoon feeding routine as the light faded, or finishing it in the dark.

Nightfall is also when large carnivores are often more active, and corrals – or stockades or bomas – can be used to confine herds and flocks during the hours of darkness. They can be effective,[12] and are promoted extensively by snow leopard conservationists across Central Asia.[13] Yet they also have their downsides. Should a predator, especially a cat, get into a corral like this, frenzy killing, in which large numbers of livestock are killed or injured instinctively, can occur. The use of corrals also requires significant changes to husbandry methods that for decades, or even centuries, have relied on livestock being left in fields or on hillsides and checked once or twice a day.

Like corrals, collars function as barriers to predation, but, in this case, at the kill stage of the sequence. They have been shown to work well with cats – as they often target the throat.[14] In Switzerland, as wolves more often attack the rear of an animal, they haven't been used as much, but they are more common in other parts of Europe. The biggest downside of fences, corrals and collars, due to their inanimate nature, is their limited focus on disrupting only one part of a predation sequence. Yet technology, in the form of sensors, cloud computing and AI, offers opportunities to make these 'dumb' technologies significantly smarter. An internet of living things, when coupled with sensors on and in livestock, and perhaps even predators themselves. The coming decade is likely to witness significant developments in this area.

But livestock guarding animals, and people, are already smart – except when they do stupid things – and, in theory, better able to adaptively deter predators in multiple ways. But as we've seen from David's experience of livestock protection dogs, they also come with their own pros and cons. In addition to the issues we encountered with people and their pets, several studies have noted the tendency of guarding dogs to chase and even hunt wildlife.[15] Other guardian animals utilised around the world include llamas and donkeys, both known for their mean dispositions. Spitting breakfast might be enough to deter humans, and maybe even solitary cats. But llamas and donkeys struggle with wolves, and, against a pack, can even end up on the menu themselves.

People, though, are the ultimate predators. How does our role as shepherds, including a wider focus on livestock husbandry, compare with these other deterrence methods? I travelled elsewhere to find out.

At home on the range

We bounced across the pasture. Above the roar of the engine, Ellery gesticulated and shouted to make herself heard. Two cans of bear spray sat on the dashboard of the side-by-side's open cockpit, ever present, ever ready. We were in grizzly country.

My heart stood still, but not because of bears.

Around us the golden fire of autumn swept through the aspen. A conflagration of colour, accentuated by the white of their slender trunks. The entire valley blazed yellow in a band of temperate glory. A crown of creation. Above the aspen, dark evergreens brooded. Above them still, sombre peaks reached to the heavens. The first snow of the season dusted their shoulders. My heart stood still.

I was in the Tom Miner Basin, a valley across from the north entrance to Yellowstone National Park. Here was one of the highest concentrations of grizzly bears in the USA, outside Alaska. Here too was Anderson Ranch. The multigenerational rancher Malou Anderson Ramirez had welcomed me to her home earlier that afternoon. Deeply philosophical, she had given me a whistle-stop tour of their land, explaining how – and why – they farmed alongside large carnivores.

The ranch was like a living laboratory for coexistence. They were constantly experimenting with various deterrence methods, either by themselves, with the other ranches in the valley, or in cooperation with university, NGO and technical partners. Smart ear tags with built-in sensors that could detect a flight response from the bovine wearing them and then send a notification were being trialled. Methods like fladry – coloured ropes or ribbons tied to fencing that flutter in the breeze – and even acoustic deterrence, like loud music, tended to be used only for short periods of time, to avoid the bears becoming habituated.

'Eighties rock seems to be particularly effective,' laughed Malou.

I remarked that eighties rock would be enough to make most living things head for the hills.

But there was one deterrence method that I was particularly interested in seeing at Anderson Ranch: humans.

Malou, whom we'll meet again later, headed off on the school run, leaving me in the capable hands of range rider Ellery Vincent.

Ellery had 40 two-year-old Angus heifers to check. I offered to accompany her and she swapped the quad for the two-seater side-by-side. I commented that I'd spent half a summer at a wildlife park in Canada zipping around in one of these. Riding shotgun, I jumped out and wrestled with gates of various descriptions as we headed up the hill to the herd.

Ellery spoke about her lifelong interest in horses and bears, most recently pursued through a master's in grizzly biology. Like me she'd worked on a variety of farms and zoos to get experience. Now, this seasonal range rider role from May to October brought all her interests and skills together.

She was the range rider for several ranches in the basin and spent long hours in the saddle June to August, checking on the herds every day in the federally owned grazing concessions at higher elevation. Now the herds had moved down to the valley floor for the autumn, and, in the case of Anderson Ranch, would soon leave for the winter.

Ellery defined a range rider as 'part stockperson, part wildlife ranger'. Someone who could harness their knowledge of livestock, carnivores and the landscape to enable coexistence between all its elements. Someone whose very presence, and intimate knowledge of processes pastoral and biological, deterred predators. A fusion of wild and domestic.

The association of several ranches in the valley received funding from various sources to fund the range rider programme, including state and federal government, as well as conservation NGOs. They tried to maintain a balanced diversity of funding sources, as befitted their involvement in both large carnivore conservation and cattle ranching. As well as daily checks on the cattle in the summer season, the range riders were also responsible for calling in incidents of cattle being lost to bears, or occasionally wolves, so that the kills could be verified and compensation claimed from the state of Montana.

I asked Ellery about the risks of riding range in grizzly bear central.

She felt comfortable around bears, she said, but being in dense cover made her nervous – due to the potential of surprising mothers with cubs or a bear on a carcass. Even on horseback, a bear could outrun her over a short distance.

'The best way to deal with a bear attack is to never have one in the first place,' she noted.

Ellery was about to start her PhD in grizzly bear biology, and I was struck by how her deep understanding of bear biology made her a better range rider. A cowgirl from urban California with an advanced biology degree, about to start another, she was stereotype-busting at its very best. Part of me wished I'd heard about this programme when I was an undergraduate. I'd have signed up in a flash. I'd have even learned to ride a horse for it.

Back on the hillside, the water trough was full and the heifers were all accounted for. While they were calmly grazing and barely bothering to acknowledge us, I asked Ellery how she would know if something had happened to them.

'They'd be spooked and scattered over a wide area. Agitated. Nervous.'

The temperature was falling, as the mist rolled down. It swallowed us whole and the aspens were gone. The golden circlet upon Tom Miner's brow was dimmed. An autumn chill set in and touched my bones as we rattled back down the hill.

The presence of people deters predators. Back in Switzerland, a programme to fund new shepherd huts in the Alpine meadows was being implemented,[16] as well as another that partners volunteers with shepherds to help them with fencing in the summer season.[17] In western Wyoming, on a ranch we'll visit in Chapter 8, Peruvian shepherds provided deterrence out on the range, following generations of Basques who had performed this role before them. Across the world, and throughout history, we have been shepherds. It is part of our heritage. Part of that age-old story of people versus predator. Those old, old memories that belong to all of us.

But it is expensive. In agriculture, the question of labour – its quantity, quality, availability and cost – is always a defining issue. In the next chapter, we'll explore financial tools for coexisting with carnivores, including paying for deterrence methods. For now, a couple of themes stand out.

Firstly, it is easy to sit in an urban armchair and proclaim that we can simply pay farmers to fence more and fence better as a means to coexist with carnivores. The same applies to any of the other deterrence tools we have discussed in this chapter, for that matter. Yet we are all creatures of habit and begrudge changes to our precious and finely balanced daily routines. This is especially the case when, in Switzerland and in much of Western Europe, we haven't had to worry about farming with large carnivores for a very long time. We're not just talking, then, about minor changes to daily routines but about major amendments to entire ways of life. It's no wonder that this issue is a contentious one for farming communities across much of the continent.

Secondly, it is also easy to sit in an urban armchair and note that the prevailing economic and policy headwinds are making farming in upland and marginal areas less financially viable, and therefore ripe for replacement by alternative rural industries, such as rewilding. That certainly seems to be the direction of travel in much of Europe, partly due to the land sparing/sharing debate that we encountered in the previous chapter – more intensive production in the fertile lowlands can free up land elsewhere for nature. Yet spare a thought for the families and communities at the centre of this shift. Systemic changes like this, just like disruption to our daily routine, are never easy.

And while the return of large carnivores may not be the cause of these systemic shifts, they are the cause of the husbandry changes required to manage coexistence with them. Given their symbolic nature, these predatory species can quickly become lightning rods for discontent by those most affected by these changes, whether at the micro or macro level. The additional burdens of time, energy and labour required to implement protection methods for livestock can be a source of stress for farmers, not to mention the trauma of losing beloved and valuable livestock to predators. In Switzerland, the Swiss Farmers' Union regional branch in Graubünden, when I visited, were about to pilot a peer-to-peer support network for farmers who had been negatively affected by such losses.

This story of competing with our fellow predators for resources is as old as our species. It is ingrained in our psyche. It is ingrained in our memories.

But for the last several centuries in most parts of Western Europe, we have had the luxury of forgetting. No longer. Wolves, lynx and even bears are back and here to stay across the region. Farming with carnivores is a reality once more. The challenge is to find effective management tools that simultaneously minimise the risk to livestock and the hassle to farmers.

In Britain and Ireland, deterrence is but one of the suite of factors that we must consider as we debate the reintroduction of these animals to our islands. It is notably easier for lynx than for wolves or bears. Yet due to the symbolic nature of all these animals, the real and perceived risk to livestock, and the real and perceived cost and effort required to adapt husbandry practices to their return, the methods acquire a symbolic meaning in their own right.

A few simple, technical changes that farmers ought to make so that carnivores can roam our countryside again? Or the straw that breaks the camel's back for over-burdened family farms already running to stand still in a maelstrom of change?

Whether in Graubünden or Montana, Argyll or Antrim, farming with carnivores is anything but straightforward.

CHAPTER 7

Money Talks

It was a dull, damp January morning just outside the village of Putten, in the central Dutch province of Gelderland.

A neighbour rang Richard van de Wetering.

'I've just passed the field with your sheep. Something's wrong. You'd better come quickly.'

Richard jumped in his van and sped down the road from the farmhouse to the field, where they grazed the land in winter for a local dairy farmer.

Something was wrong indeed.

The flock of 50 sheep was huddled in a far corner of the pasture. Except that there weren't 50 any more. Four bodies lay on the cold ground. Blood congealed on the heavy clay.

Others showed sign of injury. Red sullied the white of their fleeces. All of them showed signs of severe stress, were near panic stricken.

A wolf had come in the night.

Richard and his wife Stefana, along with their two young children, lived on this small farm, with its picturesque traditional farmhouse. Handed down from Richard's parents, and like many similar farms across the world, it was big enough to be a full-time job but small enough to struggle with economic viability. As well as raising dairy calves for veal, they rented out some of their sheds for storage. Their small flock of sheep was part hobby, part income stream from the limited amounts of grazing land that they owned outright. But there was nothing partial about their love of this land: the van de Weterings were fully invested in this place they called home. Their roots went deep. Richard and Stefana weren't just living on or off the land; they were of the land. As much a part of it as the tree-lined rural roads; as the ubiquitous waterways of the Dutch countryside; as, even, its wolves.

As I sat around the farmhouse table with Richard and his wife Stefana almost 18 months after the incident, I could still feel and hear the emotion in their voices as they told me the tale.

My mind flitted back to my gander taken by the fox on a similar winter morning on that Co. Antrim farm. A part of me understood that the wolf was just doing what wolves do. But that knowledge seemed like cold comfort here. And a part of me strongly empathised with Richard and Stefana, understood how hard this must have been for them. I felt torn.

The wolf was probably a transient young adult, freshly excluded from its home pack and in search of a territory of its own. Like many youngsters in the same situation, it was probably still figuring out the whole independence thing, including how to kill efficiently.

Unfortunately for the van de Weterings, the young wolf's emerging abilities in this area collided with their flock of sheep. Three were already dead by the time Richard arrived on the scene. Another had to be euthanised due to the extent of its injuries. A further twelve were hurt. A neighbouring farmer, 300 metres away, lost nine sheep on the same night.

On discovering the grisly scene, Richard immediately called their vet. In turn, he contacted BIJ12, the Dutch government agency responsible for responding to situations like this. A representative from BIJ12 came out that afternoon to take DNA swabs from the dead sheep, by which time Richard and Stefana had brought the remaining animals into a barn. After visiting the second farm that had lost sheep, the official returned on the same day to value their dead livestock.

So far, so good. Despite the tragic circumstances, it seemed like a textbook, and rapid, response by the relevant government body. But it was what happened next, and what hadn't happened previously, that shocked me, and turned the van de Weterings from being ambivalent about wolves and their presence in the Dutch countryside to being implacably opposed.

In the Netherlands, DNA testing has to be carried out prior to compensation being authorised, to rule out potential contamination or confusion with livestock kills by domestic dogs. The swabs are sent to the lab in batches, so delays of six to ten weeks can occur. However, Richard and Stefana's farm was not in the official wolf risk zone, an area largely covering the central forested region of the Veluwe. It was here that the first wolf pack had established itself from 2018/19, after sightings of itinerant individuals in the country from 2012/13 onwards.

Because their farm was not in the approved risk zone for livestock predation by wolves, it took three months for the van de Weterings to be offered partial compensation. It took another three months for them to receive it. When they did eventually receive the settlement, they were unhappy with it, so they challenged the process, and the amount, in court. The lengthy legal proceedings dragged on and on. In fact, by the time I visited them in June 2023, the process was still ongoing. And they still weren't happy.

Bureaucracy is a common complaint about livestock compensation schemes. When I conducted my PhD fieldwork in the Annapurna and Everest regions of Nepal, complaints about these schemes were commonplace, either that they were functionally absent or that they were simply dysfunctional.[1] And it wasn't confined to the Nepali side of the border. On the Chinese side of Mount Everest, similar critiques had been made of the compensation scheme there.[2] Fed up with the bureaucracy, some Nepali communities in the Everest area set up their own community-based compensation schemes. These were made possible by the establishment of innovative local microfinance institutions called Savings and Credit Cooperatives,[3] supported by the Snow Leopard Conservancy, with which I was partnering for my PhD research.

In one global review of several hundred compensation schemes, 75% of the negative comments related to programme administration.[4] Back in the Netherlands, Dutch penny-pinching seems to have made a difficult situation worse. During the design of the scheme, various stakeholders advised the provincial governments to expand the area within which livestock protection and compensation were readily available beyond the official risk zone. This was to address the likelihood of a dispersing wolf, like the probable protagonist in this story, killing livestock beyond the core area. Due to fear of the financial cost, this sound advice was ignored, although a raft of reforms were later introduced in 2023 to make support for both deterrence and compensation more useful and accessible.[5] But it came too late for the van de Weterings.

Back around their kitchen table, over mugs of strong coffee, the conversation continued.

'What has the impact of all of this been on you?' I asked the couple.

Richard and Stefana described the dual emotional and economic toll the process had taken on them. Aside from the shock of witnessing the damage to their small flock, many of whom they had raised from lambs, they were now questioning whether they had a future in sheep farming at all. Other sheep farmers they knew in the area, especially hobby ones, were also thinking of getting rid of their sheep. The van de Weterings just weren't sure what would replace this income source for them, although the main focus of their farming operation was raising calves indoors rather than sheep outdoors.

We headed outside to see the flock. Richard's clogs clattered on the concrete as we crossed the yard. Beyond a barn were the remaining animals. On spotting Richard and his bucket of feed, a scrum ensued at the gate. Climbing over, he was nearly bowled off his feet as he scattered the meal for them. We laughed.

I tried to guess the variety of breeds: 'That's a Texel, by the looks of it? And maybe a Rouge?'

Stefana continued: 'For six months after the wolf attack, the sheep were very skittish. Not only towards people, including us, but especially of dogs.'

But it wasn't just their sheep that were impacted by the events of January 2022. Richard's voice thickened as he spoke: 'Before the attack, I taught my children not to be afraid of wolves. After the attack, our four-year-old daughter became very afraid of the wolves attacking the sheep again, or even of entering our house.'

It was clear that the incident, its aftermath and the ensuing financial and legal shenanigans had taken their toll on the family.

What could have been done better? I wondered, as we headed out to see the field where it had all begun.

The pasture, when we reached it, seemed anything but a killing ground. Buttercups thronged the verges. A blue sky beckoned overhead, as the June sun beat down on us. Part of the meadow had recently been cut for the first crop of silage; another part had just been tilled. Verdant green woodland ringed the space on all sides.

But it was this picturesque and lovely woodland that had been deadly for the sheep. It provided cover for the predator, as forested verges like this do around the world. Recognising the elevated risk, Richard and Stefana had learned on their feet. They'd stopped using this particular area for grazing. They'd also experimented with makeshift night-time corrals, but found them to be time consuming for them, and unhygienic for the sheep.

I asked if it would have helped if they had been included in the risk zone and had received support to fence their animals in.

'We graze up to twenty meadows over the winter,' noted Richard, 'some of which are five to six hectares in size. It would take me a day to fence each of these with wolf-proof fencing, instead of the hour it normally does.'

'What solution would work best for Dutch farmers?' I asked.

'Not having wolves here at all,' they quipped. 'That worked very well for 150 years.'

'That, and a very big gun.'

The van de Weterings' story illustrates two important points in our journey to understand how we can coexist with carnivores when they return. Firstly, as we saw in the previous chapter, it is very challenging for farmers to adapt their husbandry practices to accommodate the new threats posed by the presence of large carnivores. These practices require investments of time and money: whose responsibility is that? Secondly, when prevention fails, and compensation is required, it is also very difficult to strike the right balance between the probity required to safeguard public funds, on one hand, and the speed and efficiency required to cover the financial losses of those affected, on the other.

In this chapter on finance, it is also worth noting the limits of finance. For there are some things that no amount of money can buy. But for the market value of four sheep, the Dutch state has lost the trust of this farming family. Wolf conservation in the Netherlands has lost the van de Weterings' acceptance of the species in the country. The entire rewilding movement has lost its legitimacy in their eyes.

What price to put on that?

In short, we must urgently ask: who pays to coexist with carnivores?

A licence to print money

Splat! Another mosquito bit the dust. Legs splayed, proboscis askew, it was plastered up against the mosquito screen on the window of my bedroom. I counted three of the little blighters either dead or dying on this window alone. And my long, narrow bedroom had three windows. I moved on to the next one. As I did so, I wasn't remotely worried about the death and destruction I was raining down from above. Nor was I worried about reducing the risk of me or my family contracting malaria, which was the ostensible reason for targeting these airborne killers. I was thinking about only one thing as the mosquitoes on the next window screen met their doom at my hand: money.

It was the rainy season in Malawi. After months of being baked dry under a scorching tropical sun, the red earth cried out for moisture. People cried out too, whether to God, the gods or the meteorologist, as they waited for the rains to water their maize seeds. On the Blantyre mission site, where Henry Henderson, a contemporary of David Livingstone, had first pitched his tent, and established what became Malawi's commercial capital in 1876, our colonial-era bungalow sweltered. My bedroom, on the northern side of the house and with a low, sloping ceiling, became like an oven beneath our metal roof. I baked.

I waited for the rains. The nation waited for the rains. The land waited for the rains.

Then came the first rumble of thunder. A sudden peal, like cathedral bells announcing the news. Raindrops fell as tears from heaven. People danced in the streets.

The earth resurrected. Where dust-devils spun over dry earthen ridges and yellowed-grass faded: behold, there was life.

The faded gold and deep, earthen red gave way to vibrant greens. The mountains of Blantyre that ringed the city – Soche, Ndirande, Michiru, Chiradzulu – groaned under their mantles of new growth. A chorus of maize seedlings poked their heads above the furrows.

I counted down the days until the rugby pitch at St Andrews International High School was soft enough to play on again. The rainy season was the only time of year we could play the sport full contact without requiring a skin graft afterwards. Goodbye the long summer term of swimming. Goodbye entire afternoons wasted at cricket.

But as pools of water gathered everywhere, mosquitoes bred. Malaria surged. And as the mighty Shire, carrying the entire overflow from Lake Malawi to the Zambezi and then on to the Indian Ocean, swelled with rainwater, the Kapichira Hydroelectric Plant sputtered. Rolling blackouts ensued. Paradoxically, despite water, water everywhere, water shortages followed. The bath in my parent's en suite was filled as a precaution.

Back in the Hanson household, the war against mosquitoes was waged without mercy. No quarter was given. Although I'd had malaria as an infant in northern Malawi in the late 1980s, not one of the six of us had since contracted it during our time in the country we came to call home, from 1999 to 2007. This wasn't by accident but by design. The windows were screened with mosquito netting, though some of the gaps were big enough for geckos, never mind insects. Every day at dusk we put on long trousers and socks, denying the airspace around our ankles to the bugs who lurked under beds and sofas. We doused ourselves in insect repellent. We slept under nets impregnated with insecticide. We took anti-malarial medication daily, at least for the first few years before concerns about its long-term health implications stopped that. The entire house smelled like a demented aromatherapist's and felt, for mosquitoes at least, like the demilitarised zone between North and South Korea.

Still saboteurs sneaked in. Our defences were infiltrated by an enemy that never slept. This rainy season was particularly bad. My parents turned to the last remaining weapon in their arsenal: they posted a bounty.

The princely sum of 50p was to be awarded for every dead mosquito. My brain did the maths. My eyes saw kwacha signs. Six or seven bounties would equal my weekly pocket money take. I went to work with a vengeance.

Splat! Back in my bedroom, I completed my morning sweep of all three windows in my bedroom. Mosquitoes often congregated on the screens at first light. I made sure I was there to meet them. Then, before my sister – and main competition – was up and about, I raced round all the other windows in the house. I rained down death. And the kwacha flowed. It was a licence to print money.

Human behaviour is easily swayed by penalties and incentives. Financial ones seem to have particularly profound effects on us. After all, when money talks, we listen.

When it comes to paying for carnivore coexistence, the premise is simple. As with the multilayered mosquito defences erected in and around the Hanson household in Blantyre, prevention is better than cure. If you totted up all the money we spent on deterring mosquitoes over the years, it would probably add up to a hefty sum. Add in all the time spent coating ourselves in insect repellent. Or the damage to my street cred from having to wear long trousers and socks to evening events and socials when all my friends remained in our standard attire of shorts and flip-flops.

Yet as each of my friends came down with malaria at one time or another, I never did. None of us Hansons ever contracted the illness over eight years in Malawi. Deterrence worked, but it came at a cost.

Protecting livestock from predators can cost considerable amounts of money. In Switzerland, three of the four million Swiss francs spent on carnivore coexistence in 2020/21 went to livestock protection methods and advice.[6] Just 172,500 francs was spent on compensation for livestock losses, split 80:20 between the federal and cantonal governments, approximately 4.3% of the total spend on carnivore coexistence. This amount has since increased to match increased wolf presence and activity in the country.[7] In the Netherlands, the amount spent on compensation for wolf losses is even less. In 2021, this was 46,093 euros, 0.13% of the total spend on wildlife damages.[8] Although compensation for losses to wolves increased significantly in 2022, including to the van de Weterings, it still pales in comparison next to the €18,806,433 spent on compensating damage from one species of goose alone. In fact, birds make up nine of the top ten species responsible for the majority of the €36,741,834 compensation paid in the country in 2021, with badgers, at number nine, being the only mammal on the shortlist.

But if you consider what the Dutch or Swiss states, or the British or Irish ones for that matter, spend on education, health, defence and infrastructure, these sums to facilitate carnivore coexistence are mere pocket change. Across Europe, for instance, the EU will spend €386.6 billion on the Common Agricultural Policy between 2021 and 2027.[9] Yet because of the highly symbolic nature of these animals that our instincts and culture have conspired to saddle us with, these financial figures for both prevention and cure, just like the damage they can cause, are refracted through a certain lens. Money well spent? Or pouring cash down the drain? An unfortunate side effect of having predators return to our landscapes? Or a threat to the future of rural life as we know it?

As we've already noted, however, money has its limits. What price to put on living a malaria-free life for eight years, especially given the risks from the disease, particularly the cerebral variety? What price to put on having large carnivores back in our landscape? What price to put on the van de Weterings' lost support for wolf presence?

In the rest of this chapter, we'll take a look at the different financial tools that can be used to pay for coexistence, including compensation, insurance and proactive payments. As well as some of the implementation challenges that we've already witnessed, we'll also consider some of the sustainability challenges – in other words, how to ensure there is money around to pay for all of this over not just years, but decades, and maybe even centuries. Thirdly, we'll consider whether financialising this relationship between predators and people poses more problems than it aims to solve. Is there a danger that their presence becomes, for some, a licence to print money? Is this necessarily a problem?

The last point is important because when money is introduced into human relationships, it can change things. We listen when money talks and alter our behaviour accordingly, whether to avoid penalties or accrue incentives.

Back in Blantyre, I was merrily slaughtering mosquitoes and minting it as I went. Then I realised something. One day, in my parents' bathroom, I looked in the bath. The same bath that was full of water so that we could flush the toilet when there were shortages. I could see microscopic movements that didn't quite break the surface tension. Insect life forming. Reproducing. The penny dropped. Mosquitoes were breeding in our bathtub.

My growing entrepreneurial zeal synchronised with my burgeoning biological knowledge. This wasn't just a licence to print money. It was a licence to breed it. My head filled with kwacha. Malaria risk be damned! I was going to get rich. Plus, as I convinced myself to salve my conscience, these were clean, in-house mosquitoes. They'd never encountered the malaria parasite in their life. How could they? The original population founders might have got into our house. But their progeny would never get out alive. I intended to see to that.

I went back to bounty hunting.

I never told my parents about my discovery. On reading this, they hopefully won't present me with a bill for reparations, pus two decades' worth of interest. I kept on killing mozzies for as long as the bounty was in place. The bathtub stayed full for the rest of that rainy season. Having a vested interest in the status quo meant I kept my mouth firmly shut. Relatively small amounts of money had shaped the incentives I faced, and shaped my behaviour in turn.

Money talked. I listened.

Splat!

The price of everything and the value of nothing

The theory of how financial tools are supposed to work to promote coexistence between people and predators is relatively simple. They do so by shifting the financial burden from those incurring costs from living alongside large carnivores to those who desire their presence and persistence in landscapes and are willing to pay for it, be they public, private or civic actors. Setting aside the issue of paying for the range of deterrence methods that we considered in the previous chapter, there are a number of approaches that can be taken.

Firstly, there is compensation. We've already considered this in some detail. It can be tied to requiring minimum livestock protection methods being in place, as in Switzerland; or not, as in the Netherlands. Sometimes a multiplier is applied to pay for additional livestock losses over and above the one killed. This assumes, based on scientific estimates, that if a single animal has been killed, then, as with the van de Weterings, a number of animals are probably killed, injured or missing. I encountered this in both Wyoming and Montana, where losses frequently occur during summer grazing on federal or state land. In these vast landscapes, finding all the missing or dead animals is like looking for needles in a haystack – even with range riders like Ellery on patrol. The benefit of the doubt, backed up by scientific evidence, is given. The exact multiplier can vary depending on the type of livestock.

In Colorado, where wolves are being reintroduced after a ballot initiative on the issue in 2020 and which we'll also visit later in the book, the state is planning to experiment with innovative compensation methods in two ways.[10] Firstly, it will offer a higher multiplier for those with some deterrence methods in place, thereby incentivising prevention of livestock losses. Secondly, it will offer compensation for secondary losses. This covers the hard-to-quantify knock-on effects on livestock's health and productivity from predator presence, a common gripe on my travels for this book. To link it back to the Dutch example earlier in the chapter, of the 46 still-surviving sheep in the van de Wetering flock, the 12 directly injured may have lost weight, say, or miscarried if they were carrying lambs. The remaining 34 may also have been affected in more subtle ways: they may have struggled to put on weight due to the stress of the incident, or may have been less likely to conceive when put to the ram later in the year. For farmers attuned to the tiny details of their animals' health, these issues matter. For farmers attuned to the tiny financial margins influenced by these details, these issues matter.

But they are hard to measure and manage. In Colorado, if a livestock owner is able to provide three years of baseline data about the weight gain or conception rates of their herd or flock, and they 'have had a confirmed wolf–livestock interaction resulting in livestock injury or death', they will be eligible for compensation of these secondary losses.[11] Complicated? Certainly. But does it offer the opportunity for organised farmers to be more fully recompensed for the costs of coexisting with carnivores? Probably, and the idea has also been proposed for crop damage by wildlife.[12] As we noted in the previous chapter, there is often a link between husbandry standards and levels of livestock loss to all sources, including predators. There is often a link in animal agriculture between measures like these – what is called benchmarking – and profitability. Here, as elsewhere, what gets measured gets managed. In other words, larger, more commercial and better-managed livestock units will be able to avail of this sort of support more easily. Smaller, hobby or semi-commercial, and less well-managed farms will find it harder.

The second financial tool is insurance. Although the outcome is the same as compensation – money back for losses incurred – the main difference is that a premium is required. Unsurprisingly, this sort of financial tool appears to be less common and less popular. Conceptually, at least, it gives the livestock owners more 'skin in the game' and incentives to protect their livestock and not submit false claims, else they lose their carnivore coexistence 'no claims bonus'. However, a lot of this lives in the theoretical realm, because I didn't encounter a livestock insurance scheme anywhere on my travels – although it has been used effectively with snow leopard conservation in Central Asia.[13] Here, it has frequently been paired with some sort of community fund, so that value is generated for the wider community and not just livestock owners.

Thirdly, there are proactive conservation payments, a sort of results-based environmental performance payment. In one example, from Sweden, Sami reindeer herders were paid for the number of successful lynx and wolverine reintroductions on their land.[14] Interestingly, the payments were made at the village level rather than at the individual level, though this may make sense in communities where nearly all households are engaged in a similar activity on a communal basis. Again, as with insurance, it's less common than compensation. And like insurance, I didn't encounter it in any of the countries I visited when researching this book. Yet schemes like this can and do work, as part of a broader approach to paying for environmental results that we'll explore in more detail in Chapter 13, on economics.

As with deterrence methods, each of these financial methods has pros and cons. In their review of the various types of financial schemes, Amy Dickman and colleagues propose a 'Payments to Encourage Coexistence' fund, which combines elements of all three.[15] This fund provides compensation or insurance to cover the costs of those losing livestock to carnivores, thereby reducing the direct financial impacts of coexistence. Conservation payments are paid directly to all land or livestock owners so that value accrues in a positive way to all those with these species on their land, helping to turn them from pitfall to windfall. And the community-based element of the fund provides grants for local services and activities, thereby ensuring that the financial benefits of having predators around reaches everyone, not just land and livestock owners.

A scheme like this would go some way to addressing two of the concerns raised in my dialogue with all five of the main farming unions across Britain and Ireland, as well as several livestock-specific organisations. Firstly, there is a worry that benefits from financial income associated with predator tourism would not trickle down to those landowners losing livestock to these predators. We'll consider this further in the chapter on enterprise. Secondly, many note that politicians and their policies, as well as NGOs and their projects, tend to run in five-year cycles. Farmers, by contrast, are on the land for decades, sometimes generations. From their point of view, the timeframe of landowning needs to be reconciled with that of policies and projects, so that financial support to pay for coexistence is in place over the long term.

In Colorado, one such approach is being taken by Colorado State University's (CSU) Wolf Conflict Reduction Fund (WCRF).[16] A joint venture between CSU's Center for Human–Carnivore Coexistence and its agricultural extension services, the fund aims to offer small grants to facilitate coexistence between wolves, people and human livelihoods as the species returns to the state. While the goal is to eventually have an endowment in place, the current budget is small, despite the passion of the people behind it and the importance of its neutral facilitator role in the middle of a heated debate between stakeholders. Yet I was struck by the disparity between the amounts of money poured into the campaign for Proposition 114 to reintroduce wolves in 2020, including by a small number of deep-pocketed donors, and the amounts currently available to the WCRF to facilitate coexistence. If only those funding the campaign to mandate wolf reintroductions to Colorado had matched their donations with money to pay for deterrence and compensation methods over the long term. There'll be more analysis of the Colorado context, which holds a number of significant lessons for the Britain and Ireland, in Chapter 10 – on governance.

Aside from the bureaucratic complaints about all such financial schemes that we've encountered already, there are other critiques. As we saw with the mosquito bounty in the Hanson household, financial incentives and penalties alter human behaviour. That, after all, is the point. But given our species' fondness for finance, this can lead to a range of unintended consequences. And some of these consequences can be more significant than letting a bunch of malaria-free mozzies breed in the bath.

Firstly, there will always be attempts to game the system, such as a minority of unscrupulous livestock owners trying to claim for more losses than were actually incurred. Or putting down losses from domestic dogs – which can't be claimed for – as losses from, say, wolves – which can be claimed for. The problem is that the greater the validation effort required for a compensation scheme to weed out the odd example of bad practice – such as the DNA testing in the Netherlands – the slower and more cumbersome the scheme becomes, leading to more potential conflict between the compensator and the compensated. This is an area where technology may be able to assist, such as via the high-tech ear tags we witnessed in Chapter 6. Automated flows and analyses of data from a range of sensors may help to strike a better balance at this intersection.

A second critique is that providing compensation without insisting on some deterrence being in place can reduce the incentive to minimise losses in the first place, something that is called moral hazard. Put another way, it can lead to the cure being favoured over prevention. This has been pointed out by one review of livestock compensation schemes across Europe.[17] However, the same study notes that only a few richer countries on the continent pay for the additional costs associated with adapting husbandry methods to carnivore coexistence. As long as that is the case, farmers – famously thrifty, often because their profit margins are so tight – will avoid shelling out time and money for these methods upfront. This underscores the point that financial tools cannot be considered separately from deterrence tools, and how these are paid for. Perhaps the Colorado experiment in paying higher compensation rates for those practising prevention will prove to be successful. Watch this space.

A final critique of compensation schemes is that they are an implicit subsidy for agriculture.[18] This is especially the case, so the argument goes, in marginal areas close to or in landscapes of high conservation value. They can maintain, or in some cases even increase, the presence and practice of otherwise unprofitable agriculture here, with knock-on negative impacts on biodiversity. This is an argument that dovetails with those of the Half-Earth movement, and to a lesser extent the land sparers, whom we met briefly at the end of Chapter 5, as well as some parts of the rewilding movement: with these forms of human land use gone from these places, nature – including large carnivores – can return and flourish. The issue of agricultural subsidies, and their role in maintaining farmers and farming in these biodiverse ecotones, or interfaces, between the 'wild' and the 'domestic' is a valid point that we will return to in the next section of the book. And public opinion, and with it policy and funding, is certainly moving in the direction of a more multipurpose countryside, with space for wildlife alongside farming and a range of other human activities.

Nor can the Zipperts' 12 cows compete with the vast dairy units of Cork or California, or the van de Weterings' flock of 50 with the colossal sheep operations of Argentina or Australia. Nevertheless, their right to farm in the places they call home must endure. On the issue of large carnivore reintroductions to Britain and Ireland, I may sit on the fence, but on this issue I do not. Farmers like these are as much a part of their landscapes as wolves, lynx and bears. Farms like these are as much a part of their landscape as the forest and the bog and the rivers that run through them. Does that mean that we simply freeze these people, places and processes in time, offsetting the productivity gap between these farms and their hyper-industrialised contemporaries through generous subsidies in place for forever and a day? No.

But as these people, places and processes change in response to the spectrum of forces shaping them, as they always have; as we all come to appreciate the need for and value of multipurpose rural landscapes, those who wish for or manage the return of large carnivores to these same landscapes should, perhaps with more and better funding for deterrence, compensation and coexistence, and definitely with more humility, seek to work with farmers like these.

The lesson for Britain and Ireland is again this: that those most affected should be the most consulted. And those most affected, if most consulted, may turn out – in some cases but not all – to be some of the most significant stewards of the landscapes of the twenty-first century, whether or not lynx, wolves and bears also call these home.

Coexisting with carnivores is not a licence to print money. In fact, it can impose significant financial costs, and these costs can be heavily concentrated in emotive one-off predation events that capture headlines and harden hearts. Nor is any financial scheme to pay for such coexistence perfect, be it compensation, insurance or proactive payments. Throughout my travels, I've been struck by the amount of work involved for the claimant, and the resulting resentment about the bureaucratic burden over and above any financial losses, as well as the time spent on implementing protection methods. Nevertheless, despite all their flaws, financial mechanisms remain a vital part of the coexistence toolkit. They allow public funding to be matched to the public's desires for various, often competing goals, from animal welfare and biodiversity conservation to climate change mitigation and the maintenance of cultural landscapes.

Money talks and we listen. But this chapter also highlights the limits of financial tools and valuations, as much as it makes the case for their important, albeit imperfect, role in paying for coexistence. We may be able to put a price on most things – like the death of four sheep in the Dutch countryside one damp, January dawn. But there is no value that can be put on the erosion and degradation of trust to which mishandling such incidents can lead.

And when both deterrence and financial management tools fail, what comes next in our suite of options? In the next chapter we investigate the use of force.

Money talks. We listen. And coexistence is a price well worth paying for.

CHAPTER 8

A Deadly Game

The bear charged. Low to the ground it came. An explosion of energy. A spring sprung.

It closed the gap between us in seconds.

'Surely not?' you ask. First bullies, then bison, now bears: Jonny's luck has finally run out, although, miraculously, he has made it through to Chapter 8 in one piece. More fool him, though, for seeking out dangerous situations with dangerous animals. And just to sell a few books. Karma and all that.

Rest easy. I was actually facing a bear. I was even being charged by one. But this was no ordinary bear. This was Robobear.

I was in Cody, eastern gateway to Yellowstone National Park. I was at the regional headquarters of Wyoming's Game and Fish Department (WGFD), responsible for wildlife management, conservation and hunting across this vast state – 10,000 km^2 larger than my native Ireland yet with a population less than Greater Belfast. Here, in the Cowboy State, the last of the Wild West lives on. Here, people and wildlife share open landscapes that stretch as far as the eye can see, and then some. Here, large carnivores roam.

And speaking of large carnivores, I was in the capable hands of the team at the WGFD responsible for their management throughout Wyoming, 24 hours a day, seven days a week, 365 days a year. Bears, both black and brown. Mountain lions. Wolves. Wolverines. And even the occasional Canadian lynx,[1] although the Americans, understandably, prefer it without the prefix. Canadians, after all, being either loved or loathed south of the 49th parallel as unarmed North Americans with health insurance, as the saying goes.

I wasn't remotely worried about bumping into almost any of these predatory species. I'd spent my high school years running through the Malawian bush with their African equivalents, like leopard and hyena, staying well out of my way. Except one. One gave me pause for thought, engendered a very healthy respect: grizzly bears. And as I intended to camp in grizzly country over the next few days, I was keen to make sure that I knew how to use my can of bear spray.

A few days earlier I'd come to Lander, 163 miles to the south. I was there to see Dan Thompson, head of the large carnivore unit. I arrived at the WGFD office early, having made good time from Evanston further south. I sat on the tailgate of my car and ate my lunch amidst the early afternoon sun of September's dying days. Across the car park, half-hidden in the dappled shade of the scrub that edged the site, mule deer browsed. Crickets chirped a requiem to summer's end. On the flagpole, the bison and seal of Wyoming's state flag fluttered weakly in the light breeze, unfurling slightly beneath the stars and stripes.

Dan and I talked at length, in an interview that underpins much of this chapter.

We finished with a quick tutorial on the use of bear spray, using a practice can of inert stuff that wouldn't knock us for six if my aim was off.

'You'll only have seconds to respond,' said Dan, 'so make sure the can is easy to draw.' I made a mental note to cut the fastener off the top of the can's holster, which was attached to my belt.

'The bear will be coming low to the ground. Spray downwards in a short burst. If it keeps coming, give it another burst until the spray is finished.'

I clumsily drew the can and fumbled with the safety clip. I sprayed downwards once. I sprayed twice. The can emptied.

Feeling somewhat comforted that I was starting to know what to do, but alarmed at the time it took me to draw and spray, I asked Dan: 'Have you ever had to use it on a bear?'

'Not on a bear,' he chuckled, 'but on a moose. Several times.'

'And did it work?'

'Yep – stopped her in her tracks!'

It was a reminder that, whether in Wyoming or the world over, for all our fear of them, large carnivores tend to pose less risk to us than large herbivores, whether wild or domestic.

But that didn't necessarily ease my discomfort at whether I would be able to use bear spray effectively with an animal bearing down on me at speed.

When I voiced these concerns to Dan, he asked: 'Would you like to meet Robobear?'

So here I was in Cody, a few days later, in the extensive yard at the back of the WGFD Regional Office.

One Mississippi.

I heard a noise behind me.

Two Mississippi.

I heard the shout: 'Bear, bear, bear!'

My right hand shot to my right hip and drew the can, as I turned to face the oncoming ursid.

Three Mississippi.

I fumbled with the safety clip. Robobear was on me. I died.

Thankfully, we were able to try again. Mark Aughton, who was showing me around, reversed the radio-controlled, all-terrain electric unit, with a plastic brown bear the size of a labrador perched on top. As with the XL Bully in Chapter 2, a part of my brain, rarely used these days, reactivated. Instinct initiated.

One Mississippi.

I heard the whine of the electric motor.

Two Mississippi.

'Bear, bear, bear!' roared Mark.

My hand shot to my right hip and drew the can, as I again turned to face the oncoming beast.

As I did so, quite unconsciously, I began to back-pedal furiously. Old muscle memory from the plains and playing fields of Africa.

Three Mississippi.

I flipped the safety. Robobear bore down on me, but my evasive manoeuvre had gained me a precious extra second. That was all I needed.

Four Mississippi.

I sprayed. Low to the ground in a long continuous burst of mace until the can was emptied. Robobear got it full in the face. It stopped her in her tracks. I lived.

Mark was impressed: 'That's the first time we've ever had anyone move backwards like that. Like a quarterback!'

Although, as Mark later pointed out, back-pedalling is by no means recommended when facing an actual bear. It's one thing to do it on smooth gravel. In rough backcountry, it's more likely to make you trip over a root or stone and end up on your back with an angry ursid bearing down on you.

Robobear is an educational tool used by the large carnivore team at events across the state. They use it to train people how to deploy bear spray as a form of hazing, using force – in this case, a negative sensory stimulus, to use official terminology – to effectively minimise the risk of a serious bear encounter. They also use it to impress upon people how little time they have to react. With real bears, you don't get a second chance.

With real bears, threatening human life is an instant death sentence in Wyoming. With livestock, the team tend to first catch, mark and move – or translocate – a bear that has preyed on domestic stock.

For mountain lions, which are not on the endangered species list, it's more of a 'one strike and you're out' approach. Both species are almost impossible to wean off livestock once they've developed a habit. For wolves, in the zone immediately around Yellowstone National Park, where most of the state's wolf population lives, a permit to shoot two wolves will be issued to the affected rancher. Elsewhere in the state, the species can be shot on sight, which is a bone of contention for some wolf conservationists.

Welcome to the complexities of using force for carnivore coexistence management. In this chapter, we'll look at why force is an essential, if more controversial, part of the toolkit. We'll also consider the options of hazing, translocating and lethal control in turn, as well as some of the ethics involved.

Issues of life and death have run through our story so far. They are ever present in this twenty-first-century fairy tale. The evolutionary crucible that fashioned our instincts. The social melting pots that shaped our stories and made our morality. The historical campaigns that rid our lands of large carnivores. The historical and contemporary campaigns to bring them back. Our role as missing links in managing ecosystems through hunting and farming. The use of deterrence to disrupt predators' attempts to kill livestock. The use of financial tools to put a value on a life when these attempts succeed.

Issues of life and death permeate them all. But the ethical and effective use of force brings them into sharper contrast. It is not a subject to be taken lightly. Nor is it a subject to be avoided.

For death is a natural part of life. A deadly game in which we are all back-pedalling furiously.

Back in Cody, I'd enjoyed my encounter with Robobear. But I'd also been chastened by the tiny sliver of time I had to respond to a charging bear. I was forced to confront my own mortality. It made me quite existential.

It reminded me of the three times I have been closest to death. The first, a fatal car accident in Malawi. The second, standing atop a 20,000-foot peak in the Indian Himalaya as the weather closed in late in the day and my climbing partner struggled with altitude. The third, standing face to face with Sonia, the allegedly tame tiger, at that Co. Antrim animal sanctuary in 2007.

Yet in each case, there was something about being close to death that made me feel alive. As you stare eternity in the face, reality beckons you back.

'Not yet,' it says, 'there is still too much to live for.'

And as the adrenaline courses through your veins, you snap back into the present.

Changed.

The colours are brighter. The air sweeter. Life that much more marvellous for having glimpsed it extinguished.

You breathe. Deeply.

You exhale. Deeply.

You live. Deeply.

Force forces us to confront these same existential issues with carnivore coexistence.

Deeply.

Cowboy country

The open road stretched to the horizon. Like a country song brought to life, the two-lane highway went on and on and on. As far as I could see behind me in the rear-view mirror. As far as I could see ahead of me through the windscreen. As it collided with the skyline, the road shimmered and danced in the heat haze.

Far to the south, across a plain broken by the watercourses that ran from them, were the mountains of north-east Utah. Towering peaks touched the heavens. Snow kissed their summits.

My heart skipped a beat. Again. These Mountain West landscapes were doing this so often to my poor ticker that I was starting to worry

I'd be left with permanent cardiac problems. My inner landscape was as full of wonder as this outer landscape was of majesty.

This was cowboy country. And I was on my way to meet a cowboy.

I was in south-western Wyoming, almost on the border with Utah. Across these massive landscapes, across millions of acres of land that took me hours to drive through, from those Utah mountains to the desert around Rock Springs, and back to Evanston near the state-line, cowboys still pushed herds of cattle and flocks of sheep. The old-fashioned way.

I met Shaun Sims and one of his Peruvian ranch hands at Exit 41 on Interstate 80 near Lyman. Which began a whistle-stop tour of the entire area, which continued late into the evening and then again the next morning.

To the Lyman ranch to drop off two rams. A Pyrenean Mountain Dog puppy lounged in the barn with a small group of sheep, beginning its training as a livestock guardian dog. To their family cabin between Lyman and Evanston, where Shaun's dad supervised the installation of modern plumbing. On the hilltops, turbines harvested the wind, as the Simses had harvested this land since before Wyoming became a state in 1890. Back to the Lyman ranch to fix the wheel of one of the giant contraptions that irrigated the hay meadows. I found myself lending a hand, but the light was fading and the part wouldn't quite slot into place. Then to Evanston, as night fell, to feast in the town's Mexican restaurant. And finally, to the Simses' homeplace.

As we went, always with one hand, and sometimes with both, Shaun pointed and gesticulated as he spoke. The cattle on that hillside. The sheep in that valley. The problem they had with coyotes here last year. The time a mountain lion had killed 20 young rams in a single night there.

The history and natural history of this place unfolded together. A landscape translated into a home.

I was strongly reminded of the summers I spent with my grandfather every year we were back from Malawi. And of times stretching back to my earliest memories. Sitting in the tractor cab. Graping silage for cattle. Rounding up sheep to be clipped. Taking animals to the abattoir or the market.

And like the Irish cowboy that he was, he too pointed and gesticulated as he spoke. Always with one hand and sometimes with both. The cattle in that field. The sheep in that meadow. The ploughing he'd done with his team of Clydesdales as a youngster here. The barley they'd cut there.

The human history and natural history of the place unfolded together. A landscape translated into a home.

Ulster. Utah. Wyoming.

Different land. Same love of the land.

Different hat. Same heart for their home.

Different ecosystem. Same dependence upon it.

Different nation. Same economic headwinds blowing change through these places. Whether its farmers and ranchers liked it or not. Whether they could do anything about it or not, save take shelter from its often bitter bite.

Different history. Same contested legacy of land as an asset and a symbol that people fought for. Dispossessed for. Died for.

A deadly game between people. And a deadly game with predators.

Across the 30,000 acres of land they owned, and across millions of acres of state and federal land that they grazed seasonally, Shaun cowboyed. Their 700-odd cows, and 9,000-odd sheep, rotated through the landscape and the year. From the spring lambing and calving pastures, close to hand, where coyotes were heavily shot and trapped. To the Utah mountains in summer, where bald and golden eagles were translocated under licence to protect growing lambs. To the arid land around Rock Springs, where winter snow on the ground provided water that was not available at any other time of year.

On horseback and by pickup truck. With livestock guarding dogs and with Peruvian shepherds, instead of the Basque ones of Shaun's youth. Following the forage and the seasons as wild herds of herbivores, and the Native Americans who hunted them, had done before over aeons.

As we drove north towards Kemmerer the next morning, to see the final piece of the Simses' cowboy puzzle, I quizzed Shaun about the harsh realities of life and work in cowboy country.

'If we didn't trap and shoot coyotes near our lambing grounds, our losses from predators would be much higher than they are,' noted Shaun. 'As it is, losses vary from year to year. We limit them but we don't avoid them entirely.'

'What about losses to bears or mountain lions?' I enquired.

'We rarely see mountain lions. Although they are here in the bluffs and crags. With black bears in the summer pastures, we have the occasional issue, but the shepherds and the dogs usually deter them.'

There were, as yet, no grizzlies and rarely any wolves in the area.

'Do you worry about their return?'

'Grizzlies and sheep are a bad combination,' Shaun observed. He cited another grazing area, further to the north, where the ranchers were given a set number of annual licences to cull brown bears that threatened their stock.

'What about the licensing process for catching and translocating eagles?' I wondered.

Shaun thought for a moment. 'It's not too bad. But I'm not sure how well it works. Eagles can fly what I can drive in a day. What's to stop them coming right back from where they're moved to?'

Lastly, I was curious how he felt about wolves returning. His dad had mentioned old tales about ranching with wolves, handed down through the generations. What would it be like to have to do this again?

'I wouldn't say I want wolves back in this area. But if they do come, we have the power to shoot them if they threaten our stock. That gives us the freedom to protect our business and our way of life.'

We stood on a hill in a narrow valley. I was due to head north to meet Dan Thompson in Lander. On one side, a joint coal mine and power plant was in its last years of operation. On the other side was the site for a new nuclear power plant. Change was a-coming to south-west Wyoming. Here and elsewhere, change was coming to this age-old practice of transhumance, 10,000 years in the making. Change was coming to cowboy country, whether in Utah or back in Ulster. The real or proposed return of large carnivores was but one symptom of a world evolving rapidly.

'You know you're like the John Dutton of Wyoming!' I joked. Shaun laughed. Part of the *Yellowstone* TV series had been filmed on a friend's ranch in nearby Utah. Shaun was envious of the shiny new barn his neighbour had got out of it, not to mention the hefty filming fee. I observed that I'd encountered outrage across the state about the blockbuster series. Not only had Wyoming been portrayed as an unscrupulous place full of unscrupulous people but Montana, which was home to only a small sliver of the national park, had effectively hijacked the Yellowstone brand. The Cowboy State had been held up by its northern neighbour.

About this and about predators, Shaun was not sentimental. But the use of force – whether shooting, trapping or translocating – was not something he took pleasure in for the sake of it. It was a way of life, just another part of the deadly game between cowboys and carnivores. It was also a matter of financial life and death, as business models and margins alike were buffeted by economic headwinds.

The visit had forced me to consider force not just as an abstract notion from a philosophy textbook or an academic paper. It was as much a part of that age-old competition between people and predators as the wonder or the fear. A matter of life and death.

I waved goodbye to Shaun as I headed off. My heart struggled again with the impact of yet more scenic vistas, as I tracked the Teton range north. My lungs laboured with the altitude. My lexicon ran out of superlatives.

The open road still stretched to the horizon. It led straight to some serious thinking on this serious matter.

A matter of life and death

Of the three approaches, hazing is the first we will consider. It's defined as 'making use of a range of deterrents including lighting, sound, odours, or non-lethal projectiles to discourage animals from a particular area or behaviour'.[2] Think of my bear spray and Robobear, but in a more proactive and systematic way. Or even the eighties rock music deployed alongside various deterrence methods by Anderson Ranch to terrify bears – and everything else – in Chapter 6. In fact, you could argue that hazing is just another form of deterrence and these methods, as with most of the methods to manage and govern coexistence that we're considering, do blend into each other across the spectrum of approaches.

But I've included hazing in this chapter as a distinct process that is more formal than deterrence, is carried out by trained professionals, and requires and receives some sort of official approval or sanction. In 2022, for instance, the BBC reported that some Dutch wolves were to be hazed with paintballs by local authorities due to a number of close encounters with people.[3] Behind the headline-grabbing story was a more nuanced tale that appeared to me to be less about conflict between people and predators, and more about conflict between people over predators. Which is the nub of coexisting with carnivores overall.

Hazing, though, suffers from the same affliction as deterrence methods. There is often a lack of evidence that the various methods work well or even work at all. In one memorable study, which stuck in my mind from the hundreds of papers I read for my PhD literature review, hazed Siberian tigers were more likely to cause repeated problems than those that were not hazed.[4] The probable explanation for this counter-intuitive finding, as the authors point out, is bias that came from a small sample of troublemaking tigers that may have developed a habit of killing livestock. As we noted earlier in the chapter, such habits, once set, are hard to break. In a more rigorous study of hazing lions in Zimbabwe, more concrete results were uncovered.[5] Livestock depredation fell after the implementation of a hazing programme. Some lions moved away from the area entirely; others changed their behaviour to become more elusive.

Aside from the efficacy of hazing, there are its ethics. How much force should be applied? In what form? And who should apply it? When I asked farmers and land managers in Switzerland and the Netherlands about being given the option of using non-lethal projectiles like paintballs or rubber bullets to deter predators, there wasn't a great deal of enthusiasm for the idea. Yet at the same time, they felt that wolves had lost too much of their fear of people. Could hazing be a means to restore this fear without resorting to shooting them dead? It's a typical academic response to call for more research as the solution to most problems. It keeps us in work. But, in this case, I feel more research is genuinely required. Not only on the relative effectiveness of the various technical methods available, but also on the social dynamics involved in their use and non-use.

But what if hazing doesn't work? Next, we look at removing problem animals – a process known as translocation – as the follow-on tool in the coexistence toolkit.

Conservation translocation – which includes reintroductions – is defined as a 'catch-all term for the deliberate movement and release of living organisms for conservation purpose'.[6] In the context of this chapter, it usually means catching and moving a problem animal somewhere else. Less commonly, it can also mean catching and moving a problem animal from the wild and into captivity.

But as Shaun hinted when discussing eagles, translocations are beset with challenges. Large carnivores often cover large distances. They often have very precise homing instincts, meaning they are adept at finding their way back to their original territory. Or, when they're moved, they end up in the existing territory of another animal of the same species, leading to competition. In general, conservation translocations suffer from low success rates, something we'll consider more in Chapter 11, on ecology. In one review of 33 projects, only about a third of the individuals studied reproduced, limiting the sustainability of the population beyond the life of the project.[7]

But can targeted translocations of problem animals be more effective, helping to minimise negative impacts from these individuals on humans and their activities? Again, the evidence is mixed or, as with hazing, sorely lacking.[8] Another review found the overall success rate to be around 42%, and the average cost per translocation to be US$3,756.[9]

In Wyoming, Dan also noted the challenges of making translocations work as a management tool. In the Netherlands and Switzerland currently, and in Britain and Ireland hypothetically – relatively small and crowded countries all – it would be difficult to find places to move problem animals. Bringing troublemaking predators into permanent captivity also raises its own logistical challenges, aside from the philosophical problem many rewilders would have with a wild animal no longer being free. Who's going to house them? Who's going to pay for their bed and board for the rest of their lives? It's an ethical minefield of quandaries and questions.

But the most complex ethics concern lethal control. Sometimes the non-lethal use of force through hazing or translocation doesn't work or isn't enough, whether this is based on scientific data or political realities. In theory, selective removal of problem animals is just another technical tool in the carnivore coexistence management toolkit. In practice, the devil is in the detail. How? When? What? Where? Why? As the detail involves explicit decisions about life and death for individual animals, it carries the emotion that topics like this evoke. Add to this the symbolic nature of large carnivores, and the equally symbolic nature of the human conflicts over them, and we have what is effectively becoming a major new front in the culture wars.

The years 2022–2024 were particularly significant markers of this in Europe. In September 2022, a pony belonging to EU Commission President Ursula von der Leyen was killed by a wolf in northern Germany.[10] In December 2022, the Swiss Parliament approved a law to make lethal control of wolves in the country more flexible,[11] after a rancorous public referendum on the same issue in 2020.[12] In January 2023, a kill order for the wolf that killed von der Leyen's pony was issued and then retracted as a legal brawl ensued over the animal's fate.[13] In April 2023, a kill order was also issued for the brown bear that killed jogger Andrea Papi in northern Italy.[14] Another round in the legal boxing ring got underway. The animal was taken into captivity.[15] In September 2023, the EU Parliament debated the relaxation of regulations for managing wolf populations, including via lethal control, and in September 2024, EU member states voted to downgrade the wolf from 'strictly protected' to 'protected'.[16] By the end of 2023, the implementation of the new lethal control legislation in two Swiss provinces, including Graubünden, was suspended as the Federal Administrative Court reviewed a legal challenge.[17] In the coming decades, conflicts like these are likely to intensify further.

Each of these cases provoked howls of protest from both conservation and animal rights groups. And, of course, counter-howls from farming and hunting groups. This seemed in contrast to the situation I encountered in Wyoming. Perhaps that's a function of the USA's greater ease and familiarity with matters of life and death in the great outdoors, through its strong hunting culture. That and easier access to firearms. It may also be a function of the state's conservative bent and an absence of urban areas greater than 60,000 people or so. Dan admitted that his job of managing coexistence with large carnivores would be much harder if Wyoming had large urban centres.

Screaming headlines aside, what does the scientific evidence have to say about lethal control as a management method? Well, unsurprisingly, it's a mixed bag. One review found that lethal control had no effect, although it noted a lack of evidence on the issue.[18] Another review concluded that it was the second most effective option, after livestock guarding animals, though its efficacy varied.[19] Another study found that lethal control could also be more expensive than the use of deterrence methods.[20]

A significant argument against lethal control of wolves especially, deployed in the European examples we've discussed, is that shooting dominant pack members will lead to the pack breaking up.[21] In turn, this will lead to more lone and dispersing wolves in the landscape, like the one that probably killed the van de Weterings' sheep in Chapter 7, driving even more livestock losses and even more irate farmers.

When I put this to Dan, he talked about the WGFD's approach of having a GPS or radio collar on an animal in every wolf pack in the state. He believes that this allows them to selectively manage problem packs and animals with greater effectiveness, limiting the chances of a pack dispersing. I even got to see the live feed from the collars when I visited the Cody office.

'But what of the ethics?' I asked Mark. 'Do you get people complaining that putting collars in every pack makes these wolves less wild?'

He smiled. 'Yes, but only from photographers.'

Mark's answer hints at some of the conclusions I have come to in writing this chapter. In matters of life and death, there are no easy answers.

Firstly, it is a topic that becomes highly polarised very quickly. Yet to all those vehemently opposed to lethal control in any shape or form, I would say this: spend some time with people like Shaun. Walk a mile in their shoes. And to all those who think that making lethal control easier and more flexible will instantly solve all their problems with large carnivores, I would say this: spend some time with people like Dan and Mark. Walk a mile in their shoes. Complexity and nuance abound in both cases. Surely a happy medium between open season and no season can be found.

Perhaps my own experience of managing a small farm – which also brings you face to face with matters of life and death – has inured me against sentimentality. Perhaps I have become too pragmatic at the ripe old age of 36. But, secondly, given that so much of coexistence with carnivores is about perception as much as reality, there is considerable merit in livestock owners perceiving that there are management options – including lethal control – available to them when there are significant problems with livestock loss. In fact, this sense of agency is one of the reasons why George Monbiot advocates the regulated hunting of reintroduced wolves in Britain.[22] No one likes to feel that they have no control over their own destiny, that the economic headwinds they face will blow them away. In other words, knowing that the culling option is available may be as empowering, for some, as actually implementing it.

Dan Thompson again: 'The lethal control option – and the large carnivore team's availability across the state 24/7, 365 – gives ranchers confidence in our management role.'

'And,' he added, 'so does having data to support our actions.'

Which leads me to my third brief conclusion: more research is genuinely needed to understand the use of hazing, translocation and lethal control as management options.

In Britain and Ireland, those who want to reintroduce large carnivores, those who oppose this, those who regulate it, and those who observe – from near or afar – will all have to grapple with the use of force in all its complexity. Few issues associated with managing these powerful animals are black and white; grey-hued trade-offs are the norm. Do the ethics of a reintroduced lynx dying from stress or illness as part of a conservation translocation differ from a reintroduced lynx dying from a gunshot as part of a lethal control intervention in the same project? Both processes can be informed by science. Both processes can be performed by trained professionals. Yet the latter is arguably much more controversial than the former. Why? In Chapter 14 we'll consider issues like these as part of the philosophical case for and against the return of large carnivores.

As fellow predators, humans using force to manage these animals is a necessary but controversial aspect of coexistence. Particularly important – yet complex – in crowded countries or regions, where competition for space among different users is more acute, the use of force necessitates careful and ethical handling and oversight, whether carried out by professional managers like Dan and Mark, or by land managers like Shaun.

So far, the management methods we've explored have been mostly reactive; they try to stop bad things from happening between humans and large carnivores. In Chapter 9, we'll consider proactive, market-led approaches for coexistence: can these transform predators from a problem into potential? Is there even a business case for their reintroduction?

Coexisting with carnivores is a deadly game.

Whether you live in cowboy country or not, using force in the management of this coexistence requires a delicate balance between multiple perspectives.

Emotionally charged, it is a matter of life and death.

CHAPTER 9

Animal Spirits

The crowd fell silent. A hush rippled through us. A momentary and collective sense of wonder stilled the boisterous group. Five hundred metres away, the objects of our fascination ambled into view. The grizzly bear sow strolled nonchalantly across the mountain meadow, as her cubs squabbled and wrestled behind her. She paused to dig for caraway roots between mouthfuls of grass.

One hundred metres away from the bear family, an Angus cow grazed the very same grass. Neither animal was remotely concerned with the other. It was a bucolic idyll. Almost biblical, like a scene straight out of Isaiah 11's famous passage on carnivore coexistence.

For those watching, including myself, this was a quasi-religious experience. Although the conversation amongst the group of tourists had picked up again, its nature had changed. In the presence of such power and grace, it had taken on a reverential tone. The hush was a hallowed one. In this great cathedral of nature, silence was the only appropriate soundtrack as we beheld the greatest show on Earth.

I was in Montana's Tom Miner Basin. Dusk's dying light cast its bright spell on the mountaintops. In this amphitheatre of awe, where the aspen blazed in all their autumn splendour, in this moment, silence was golden. Literally.

It reminded me of my first visit to Aachen cathedral. It was the site of Charlemagne's palace-chapel from the early ninth century, and I had written part of my undergraduate history thesis on the significance of the building and the persona attached to it. En route from Cambridge to Heidelberg in 2013 to take a summer class in Nepali at the latter's

South Asia Institute, we had stopped for a few hours in this charming city. Realising that historical artefacts and toddlers weren't an ideal combination, Paula had taken the kids to the park for an hour so that I could complete my pilgrimage to this special place I had spent years of my life studying. It is, to me at least, the most beautiful building in the world.

Beneath the Byzantine architecture, I sat. Silence was the only appropriate soundtrack. As light spilled into the octagonal structure above me, it illuminated the last vestiges of incense that still hung in the air. Candles flickered. The hand of history reached out and touched mine.

Back in Montana, I completed another pilgrimage. That very same morning, I'd caught a glimpse of a distant wolf in northern Yellowstone. Now, here I was seeing grizzly bears in the wild for the very first time. I had spent years of my life thinking about these animals, their meaning and their potential return to Britain and Ireland. My heart laboured under the weight of the wonder. The hand of natural history reached out and touched mine.

Malou Anderson Ramirez began her talk to the assembled crowd of wildlife tourists. As she had also explained to me on our tour of the ranch that afternoon, she spoke about the practical management tools they employed in the valley to allow cattle ranching and carnivore conservation to coexist. Ellery Vincent also pitched in to explain what a range rider was and why her role was so crucial to maintaining the delicate balance in the basin. The engaged audience of agricultural extension agents from across the USA asked technical questions about GPS-enabled livestock tags and the like.

But most of all, Malou spoke passionately about the philosophy that underpinned these practical efforts. Of the need to share the land with wildlife. Of the need to share the land with each other. Of the need for humility throughout.

'You're like a Mountain West version of Wendell Berry,' I'd remarked earlier.

She laughed and admitted that this wasn't the first time the parallel had been drawn.

'Him and Aldo Leopold!'

Like Berry, Malou questioned the status quo and lived its alternative. Like Leopold, she thought like the mountains she lived amongst.

But philosophy alone is not enough to pay the bills and put food on the table. Malou was also a wildlife tourism entrepreneur.

Wildlife tourism is 'the experiencing of wildlife by tourists', as one straightforward definition puts it.[1]

As important as books and films are for connecting us to other species, people and places, there's no substitute for experiencing them first-hand. Like my visit to Aachen cathedral, or the group tour in the Tom Miner Basin, that experience can be profound.

And for profound experiences that provide us with authentic connection to the world around us, people are willing to pay top dollar. In fact, in an age of artificial intelligence, authentic connections like these will become more, not less, valuable.

Money talks in wildlife tourism. By one estimate, grey wolf-related spending in the Greater Yellowstone Ecosystem adds over US$45 million to the local economy every year.[2] Given the crowds of wolf-watchers in the national park's Lamar valley that I'd encountered that morning, I could well believe it. Grizzly bear-related tourism generates millions more, though there is much overlap between the two.[3]

We put our money where our mouth is and pay for the things that we value. Our animal spirits roar. And, increasingly, we value having large carnivore remaining in or even returning to our landscapes. But as we've seen in the preceding chapters, there is often a cost attached to this, and it is often borne by those least keen on the presence or return of these species. Also in these chapters we've considered how deterrence, finance and force can be used to manage the risks associated with reconciling livestock farming and predator conservation. In summary, these are reactive methods that try to prevent bad things from happening, and then try to fix them when they, inevitably, do.

By contrast, enterprise methods see potential from the presence of predators, not just problems. They see opportunities to generate income, mainly through tourism but also, as we'll consider later in the chapter, from hunting and even from predator-friendly livestock certification schemes. And all of them work by trying to put a value on large carnivores, in the process turning them from liabilities into assets.

Both wildlife tourism and hunting create 'use' values.[4] Funds can be generated directly from using the animals, whether they are shot with cameras or guns. On the other hand, 'non-use' values focus on more intrinsic valuations. An existence value refers to the benefits derived from knowing a species like a wolf or grizzly exists, even if never 'used' in any way. A bequest value comes from appreciating that such species will endure into the future for successive generations to enjoy.

And though the 'use' value from the small crowd of tourists in the Tom Miner Basin observing this mother and cubs can be calculated based on the fees they paid for Malou's talk, 'non-use' values are much harder, if not impossible, to quantify or qualify. How to value the sense of wonder that stilled our lips and hearts as we gazed on this particular bear family in this specific Montana pasture? How to put a price on a lifelong dream of a pilgrimage fulfilled? 'Non-use' values can be powerful motivators outside tourism as well. I've spent a considerable amount of my life, career and energy researching, and being involved in, the conservation of a species – the snow leopard – that I've never even seen in the wild. Yet. Or maybe I never will.

As important as they are, though, 'non-use' values, like Malou's powerful philosophical musings, don't pay the bills directly. So surely an approach, like tourism, that allows direct value to be captured and revenue to be generated from the presence of predators, and is less morally fraught and contentious than hunting them, is an unalloyed good? In principle, yes. In practice, there are a number of factors that need to be considered.

Firstly, wildlife tourism can be a victim of its own success. Outside of San Francisco, the traffic in Yellowstone was among the most congested I encountered on my entire 2,700-mile USA research road trip. Wherever there was wildlife, traffic jams ensued. And I was there in early October. What it's like in July or August I dread to think. At the same time, roads give easy access to the park for people who otherwise couldn't or wouldn't experience places or species like this.

And away from the main trunk roads, nature still reigns supreme, as we'll discover in Chapter 11.

In the Tom Miner Basin, word had slowly filtered out that the spot we were standing on was a superb – and free – viewing spot for bears. Over recent years the number of tourists had steadily increased. The association of ranches in the valley, the Tom Miner Basin Association (TMBA), put up a sign to educate tourists about bear biology, conservation and safety, and to elicit donations. Range riders like Ellery were even tasked with educating visitors about safely and respectfully engaging with grizzlies.

Because, like livestock, we humans have a tendency to do dumb things. I heard one story of a group that set up a full commercial kitchen at the viewing spot in order to cater for a corporate function. Under the Montana sky, a full steak dinner was served to patrons. Never mind that they were in the middle of a valley with one of the highest densities of grizzly bears in North America. Never mind that this species is, along with the polar bear, the largest land carnivore in existence. Never mind that this large land carnivore can outrun a horse over a short distance and has one of the most powerful senses of smell in the entire animal kingdom. Never mind all of that.

Thankfully, no grizzlies crashed the party. Otherwise and inadvertently, there may have been a different cut of meat on the menu entirely. But it goes to show that common sense is not that common. And with wildlife tourism, especially where large carnivores are involved, common sense is needed to protect all involved, people and predator alike.

Secondly, wildlife tourism needs to be directly connected to those paying the price for coexisting with carnivores. A frequent refrain when I discussed the potential of tourism connected to carnivore reintroductions with farming unions in Britain and Ireland was that any such income generation needed to directly benefit those whose land the animals were on, and especially those losing livestock to the species in question. It's a fair point. It's also part of the coexistence fund proposed by Amy Dickman and colleagues that we discussed in Chapter 7. But making that happen in practice is complicated.

Nevertheless, in Montana, the TMBA has created some sort of shared value from grizzly tourism by establishing the organisation in the first place. As well as providing a degree of collective herd management, through shared funding of range rider positions like Ellery's, and a donation box at the viewing site, the association also charged fees for talks to tourist groups that Malou regularly delivered year-round. This didn't preclude individual enterprise either. Malou's family run a reconciliation and dialogue facility and retreat on their ranch. Surrounded by multi-strand bear-proof electric fencing, the tipi circle was a place where groups could come and stay as they ironed out tough issues together. Maybe being surrounded by grizzly bears while staying in tents helps focus minds and makes other matters seem less consequential by comparison.

Speaking of tents, at the very end of the road in the valley was a US Forestry Service campsite. I'd driven up there earlier to poke around. En route, an impressively antlered bull elk crossed the road in front of me. Just round the corner a coyote – or was it a wolf? – dashed across the same track. This place was alive. Bursting at the seams with nature in all its splendour. It was better than Yellowstone.

I parked my car and went to inspect one of the grizzly-proof bear bins placed adjacent to each camp spot. I'd seen black bear-proof bins in Canada before, but these were on a whole different level of durability. And I remembered the camping stories my dad had told me from the North American leg of his round-the-world cycle in the early eighties. Then, he and his cycling partner John Rogers had to pack their food into a bin liner and hang it from a tree. A faded photo depicted Dad posing with his heavily laden bicycle next to a sign that said: 'Welcome to Yellowstone National Park. Beware of wildlife. Stay in car.'

When I'd arrived at Malou's earlier that day, I'd enquired if the campsite was still open. It was but she cautioned against camping in it alone, especially at this time of year when I'd probably be the only person there. A part of me that was afraid to camp in grizzly country breathed a sigh of relief as she offered me the use of their quaint schoolhouse tourist rental instead. Isabella Tree's *Wilding* sat on the coffee table, alongside the *National Geographic* Yellowstone special edition from 2020 in which I'd first read of Anderson Ranch. They were about to shut down the building for the coming winter, so I shuttled back and forth to the pump in the yard for water. The visitors' book detailed satisfied visitors from across North America and beyond. If ever I write another book, and need a quiet place to do so, I think I'd pick that schoolhouse.

Wildlife tourism is never perfect. There are always trade-offs between conservation, animal welfare, profitability and visitor satisfaction.[5] Nevertheless, it is among the most powerful practical methods for maintaining large carnivores in landscapes, and for funding the costs of coexisting with them. Whether or not this is enough to justify, and pay for, the return of wolves, lynx and bears to Britain and Ireland is a topic we will assess in Chapter 13.

Back at the viewing point in the Tom Miner Basin, we could see a huge male grizzly bear stroll out of a distant aspen grove and into the mountain meadow. Another hush rippled through the crowd of wildlife tourists. Dusk fell. On this hallowed ground, silence was again the only appropriate soundtrack. On this fine autumn evening, in this amphitheatre of awe, silence was golden indeed.

But what if the 'using' of animals like this was with guns rather than cameras? I journeyed on to find out what that involved.

We're going on a bear hunt

'I fell in love with them as a species.'

I was sitting in a coffee shop in Cody, Wyoming, with Joe Kondelis. Eloquent and ebullient, Joe expounded on his love of bears, particularly black ones. In fact, his love of bears had led him to set up a non-profit called the American Bear Foundation (ABF). ABF advocated for bear conservation across the US, but especially in the landscapes of the Mountain West.

But despite being a passionate bear conservationist, Joe was not a scientist. Or a photographer. Or even an operator of bear safaris for tourists.

Joe was a hunter.

I had initially been put in touch with Joe to see if we could arrange my participation in a large carnivore hunt of some kind. I like to get under the skin of the places I visit, and the processes I study. I couldn't think of a better way to understand the many facets of hunting as an enterprise tool for large carnivore conservation. But it was a little outside my comfort zone. I had mixed feelings.

I began my career in large carnivore conservation at a young age. Enthralled and obsessed with big cats especially from as far back as I can remember, I naturally gravitated towards a focus on individual animals. An interest in animal welfare followed. Then someone gave me the 'Children's Guide to Animal Rights', or I bought it at a school book fair. It ignited a passion.

Soon I was organising a petition to ban bullfighting in Spain. I even sent it off to the Spanish ambassador in Dublin, but he never replied. Next I organised a debate on animals in circuses for my primary school classroom. Given that I hated public speaking and performance more than anything else throughout my adolescence, this was an indication of how strongly I felt about the issue. Ballybay Central School, with its entire student body of 60 across eight year groups, didn't know what had hit it.

Meat consumption was next on my list of wrongs to right. I trialled vegetarianism for Lent one year. My parents were bemused. My grandparents were horrified.

My interest in animals as individuals continued all the way through to early adulthood. In fact, I taught myself ethology, or the science of animal behaviour, and then designed and conducted that olfactory – or smell-based – enrichment experiment with Sonia the tiger and the pit bulls, precisely because I was thinking of transitioning to study animal welfare and behaviour after my history degree.

But in parallel, as my knowledge increased, I began to see individual animals in the context of their populations and the ecosystems of which they were a part. My interest in conservation began to develop. As my worldview expanded further still, I began to join up the dots with economic and social issues too. A passion for sustainability was born.

At the end of my history degree, I didn't do a master's in animal welfare and behaviour at Queen's University Belfast after all. Nor did I plump for the ecological management and conservation biology programme also on offer in the School of Biological Sciences. Instead, I went to Queen's University Management School to study business management and sustainability.

I had become convinced that reconciling ecology and economy – both, as we've already noted, from the same Greek root word, *oikos*, meaning 'home' – was not just the greatest task of our time, but the greatest task of all time. Synchronising the two realms could just about be the single most important conservation tool ever, at least in theory. I am still convinced of this. I had also come to believe that, at a practical level, enterprise could be a driving force in conservation, complementing more traditional public and philanthropic approaches. I am still convinced of this as well.

And in the tension between the rights of the individual and the needs of the population, at least in the animal kingdom, I had come down on the side of the fence that favoured the latter. The utter revulsion at hunting that I'd held so strongly as a child had given way to a curiosity about its potential as a coexistence tool. I was going on a bear hunt. And I was scared about what I might feel about it.

Except I wasn't. Despite my enthusiasm, and the podcasts I listened to in preparation, it was the wrong time of year for bear hunting in Wyoming. Joe invited me on his moose hunt instead but, as with Jaden and his elk in Lander, he bagged it just before I arrived. Which was just as well, given that he'd waited 16 years for a moose hunting tag to be randomly allocated to him. My lack of hunting bushcraft, and my inability to hit barn doors at point-blank range, might have ruined his chances for the next 16.

'It was super bittersweet,' observed Joe, on finally getting his moose.

The hunting of large carnivores remains a bittersweet topic for me, as it does for many. Yet I cannot deny the genuine passion for the process, and love for both the landscape and the species, that many hunters like Joe exude. I don't just not deny it. I recognise and accept it. I feel the same. So what if some conservationists hunt with guns and some with cameras, provided the right mix of penalties and incentives to steward and safeguard the process are in place? After all, modern conservation had its origins among the patrician hunters of Western Europe as they became concerned about dwindling game populations in Africa especially.[6]

And arguably, for those who favour the rights of the individual animal over the population, when this very same argument is applied to our species, it gives freedom to individual human animals to exercise their choice to hunt within legal boundaries. A few paragraphs, however, aren't going to do justice to the complex moral quandary this topic represents. Along with the issue of lethal control in the previous chapter, it is among the most controversial aspects of large carnivore conservation. Denying or ignoring that won't make it go away as an emotive topic that gets people fired up.

'Taking life is more profound an issue than ever before,' acknowledges Joe.

Some people like hunting. Some don't. Some people accept hunting as a reality. Others don't. But hunting is going to continue as a pastime, a necessity and a way of life, as it has since we first evolved on the plains of Africa. It really is the oldest profession.

And there is another example from human history that can help us navigate this tension as we consider the economic potential of hunting, including in relation to large carnivore reintroductions to Britain and Ireland. Whether concerning people or other animals, the implicit tension here is between freedom and equality. A population in which all individuals are treated exactly the same will not be free. A population in which all individuals are free to do exactly as they wish will not be equal. On the messy societal spectrum in between – on any given issue, not just hunting – this balance will exist in constant tension. For that reason, the fathers and mothers of the French Revolution added a third ideal to their rallying cry: fraternity.

In the debate about hunting as a tool for coexisting with large carnivores, I think we should all add fraternity to our rallying cries.

Unsurprisingly, as with lethal control, the evidence about the positive and negative impacts of hunting, from both an ecological[7] and an economic[8] perspective, is also contested. As with use and non-use values in tourism, much of the debate here is also about value. What value is in the horde of tourists in the Lamar valley photographing that wolf, compared to a wolf that is shot as a trophy outside the park? And what is the economic value of harvesting predators like these versus the ecological value of leaving them where they are? Can all this value even be measured by quantitative economic or ecological metrics alone? Some of this may be objective but as much is subjective. A matter of values more than value.

As I chatted to Joe, I was surprised at the relative cheapness of hunting tags for predators compared to herbivores. Of the US$15,000 he'd recently spent on a grizzly hunt in Alaska, Joe had spent only $200 on the permit, with the rest going mostly to a local guide.

'What of eating large carnivores as a rationale for hunting them?' I asked.

'It's important to honour the animal by eating it all,' noted Joe, 'but eating is more of a by-product, though it is often used to rationalise hunting.'

He continued: 'The older I get, the more hunting is about mental health; it's enjoying the wilderness; it's a challenge; it's primal; it's in our DNA.'

Perhaps there wasn't a million miles between Joe's hunting pilgrimages to the boondocks and my wolf pilgrimage to Yellowstone, my grizzly pilgrimage to Tom Miner or even my Charlemagne pilgrimage to Aachen cathedral.

If, through reintroductions, rewilders want parts of Europe to resemble parts of Africa and North America, with their assemblages of large herbivores and large carnivores, will they also accept the hunting that is an accepted and common activity in many of these places? Or in Britain and Ireland, where hunting is more controversial and, rightly or wrongly, often tarred with perceptions of class and elitism, would that just add fuel to the fire of an already combustible subject?

Yet support for this can be found in unexpected places. While it may have been the Lynx UK Trust's failed reintroduction attempt that inspired me to turn my attention towards this topic, it was a certain George Monbiot who inspired me to try to go on a bear hunt. Writing in *Feral*, Monbiot suggested that hunting reintroduced wolves in Britain could be beneficial, despite his personal dislike of the process, because it would create a constituency of support for their presence, provide reassurance that the species was under some degree of control, and maintain their fear of people. The prophet spoke. My pilgrimage followed.

Back in that Cody cafe, Joe continued to wax lyrical about his passion for bears, a Teddy Roosevelt figure combining hunting, conservation, philosophy and philanthropy.

'Someday I may hunt and just take a picture. Because we're having an impact, we have to take responsibility.'

Hunting and tourism trade on people experiencing – and using – large carnivores first-hand. For the millions and billions who will never get that opportunity, but who still want to connect with and support the conservation of these species through their purchases rather than donations, is there a way to harness the formidable power of the market to this end? In short, is there a way to turn lynx, wolves and bears into conservation brands of sorts? I kept journeying to discover more.

Nature calls

In Belgrade, Montana, I met the world's best compost toilet. In the rustic wood-panelled interior, there was electricity and a mirror. Art adorned the walls. Compared to some of the horrors I've encountered, mainly on community farms, this was the last word in compost toilet luxury.

I was so impressed by it that I took a picture of it. I told Becky Weed I'd give it a shout-out in my book. It was epic.

Despite what you might think, I'm not actually an aficionado of compost toilets. But if you are, it might even be worth a pilgrimage the next time you're in Montana.

I was visiting Becky and her husband Dave at Thirteen Mile Lamb and Wool to learn more about predator-friendly livestock certification. It was part of my exploration of whether the market could deliver coexistence with carnivores more efficiently and effectively than the state or civil society. Economically, could tourism, hunting and certification be used to balance the scales between those wanting predator presence and those paying for it? And philosophically – a line of thought that also seemed to exist in the entrepreneurs I was meeting – could it create a stronger and more direct connection between affected livestock producers and sympathetic consumers?

Nothing represents this opportunity more than predator-friendly certification. It involves consumers paying a premium for livestock products, like lamb and wool, produced in a way that facilitates coexistence with carnivores, mainly via the non-lethal deterrence methods we've already considered. Citizens who care about the presence of predators in a landscape pay for these goods, creating an incentive for the landowner to have these wild species around. The idea combines my loves of conservation, agriculture and enterprise perfectly. But at the same time, nothing represents the challenges of reconciling these three themes more than predator-friendly certification.

Becky explained as we walked her farm together.

She had been an early pioneer of the concept in the early 1990s. As one of the first ranchers to pilot the idea, she had a front-row seat to the potential and the problems. Starting with blankets initially, she then had some success selling hats in Yellowstone gift stores. Next she tried to sell to some sympathetic outdoor brands, but it didn't work out. The challenges and costs of integrating relatively small amounts of predator-friendly wool into vast multinational supply chains were prohibitive. Instead, Becky started her own on-farm mill to control quality and costs. They tried to create a cooperative of six wool producers but, as these were spread between western Idaho and eastern Montana, the logistics were challenging.

Another challenge emerged. For some ranchers, having 'predator' in the brand name just wasn't acceptable. 'Predator-friendly' might have gone down well on the American littoral, but not in the American heartland. Nor did the scheme's contractual requirement that lethal control not be used sit well with those who wanted to retain that option.

'Ironically,' said Becky, who served on the state's Livestock Board in the 2000s, 'a lot of Montana ranchers were starting to become predator-friendly in practice, through implementing deterrence methods, but not in name.' For her, the greatest contribution of 'predator-friendly' was as an educational tool to highlight the potential benefits of introducing coexistence methods for both small- and large-scale sheep producers.

Small-scale wool producers, though, were attracted to the growing marque, as it helped with direct marketing to customers, dovetailed with their eco-friendly branding, and added value and margin. Yet Becky, especially via her experience of establishing her own wool mill, found that it was very difficult, not to mention costly, to establish a parallel supply chain for predator-friendly wool.

I remarked that this cut to the heart of the challenges with certification schemes in general.[9] The more verification that is involved, the more trust consumers place in the scheme's ability to deliver its goals, in this case carnivore conservation. But the more verification required, the more costs and hassle involved for busy farmers, and therefore the higher the barriers to enter such programmes.

'Where does the balance lie? And can the benefits outweigh the costs?' I asked Becky.

She paused for thought. 'There is a place in the marketplace for special projects like predator-friendly. But if we want to change agriculture and food systems, we need to lower barriers to entry for farmers.'

This was something she and Dave were now transitioning to on their lovely farm. The mill had been taken over by a former employee. A two-acre field and shed now housed a small but growing market garden serving the consumers of nearby Bozeman.

Becky continued thoughtfully: 'When we think about coexistence, about predator-friendly, we also need to zoom out to consider the bigger market and state forces that shape agri-food systems.'

'But nature,' she concluded, 'is the ultimate arbiter.'

The lights of Bozeman in the distance twinkled as the daylight began to fade. We headed back to the farmhouse.

This was a piece of land that was deeply loved. And deeply cared for. Down to the art in the compost toilet. Or the pebbled ramp for the occasional visiting cattle herd to drink from the creek without trampling the bank. It was an outward manifestation of Dave and Becky's inner belief that ecology and economy, like agriculture and conservation, should synchronise outwardly. The implicit made explicit in this place they called home.

Predator-friendly certification is an example of how this can work, as well as of all the many challenges inherent in turning bold ideas into viable businesses. It symbolises the potential as well as the pitfalls. Although Becky and Dave are no longer predator-friendly certified, the work they helped pioneer is now carried on by the Wildlife Friendly Enterprise Network (WFEN).[10] From those humble beginnings in the 1990s, WFEN continues to pioneer predator-friendly certification in North and South America, as well as similar wildlife-friendly certification programmes involving a range of species across the world.

But this example also shows how, in a world where cheap fossil fuels have underpinned enterprise, and civilisation, for a quarter of a millennium, it is very difficult to switch to a more circular economic model that synchronises ecology and economy. In toilet terms, we're flushing diversity down the drain. The linear model is taking, making and breaking the planet. In toilet terms, we need a change to a compost loo. A circular model that takes, makes and regenerates the Earth that is our one and only home.

Becky and Dave's compost toilet is not just the best in the world. It is a symbol of the world. Or, more precisely, the world as it could be. Should be. Must be.

For that to happen, to safeguard and sustain all life on Earth, the right set of penalties and incentives are needed. As someone who suffers from a chronic, and occasionally acute, form of optimism bias, which I need to treat with daily doses of reality, my natural bent is towards carrots. But we do need sticks as well.

Because if we don't get it right, nature is going to arbitrate all these things for us.

I hope tourism, hunting and certification schemes that facilitate carnivore coexistence can all be part of this new regenerative economy being slowly and painfully birthed into existence. For as humans, we are engines of enterprise. Our insatiable thirst for new economic opportunities cannot be quenched. It is a hallmark of our race. Our animal spirits, in the best sense of the term. People should therefore be encouraged to maximise the commercial opportunities from potential large carnivore reintroductions, as well as from where these animals naturally recolonise, and where they have always been.

Overall, the market plays a role complementary to the state and civil society in facilitating carnivore coexistence. It can add to the reactive methods explored in the previous chapters by creating proactive and profitable ones. But care does need to be taken to address weaknesses, and potential failures, in the various models. And to consider the ethical dilemmas that accompany tourism, hunting and certification alike.

As important as husbandry, finance, force and enterprise are as management tools, however, they don't address the clashes of vision and values that usually underpin conflicts between social groups over large carnivores. Nor do they resolve these conflicts in and of themselves, provide space for different perspectives to be recognised and reconciled, or consider the power dynamics that often lurk in the shadows in situations like these. Only governance can do that, as we will discuss in Chapter 10.

Back in Belgrade, and across the world, nature calls. May our animal spirits respond.

CHAPTER 10

Common Ground

The ecosystem bustled with wildlife. Some individuals basked in the sun of this late September day. Others rested in the shade. Still others gathered in and around the numerous watering holes.

The wildlife itself was not hard to spot. Every conceivable variety was on display. Every pattern of pelage. Every phenotype.

I watched all of this under the magnificent spread of an old red oak. Its girth spoke of a long legacy of life. Its mantle spread out over the main game trail. Stretching down and around it in all directions was a city of trees.

But this was no ordinary forest. This was the urban jungle. This was Fort Collins, Colorado.

A lovely college town nestled against the Rocky Mountain's foothills, Fort Collins began life as a military outpost in 1864. Later, it became home to Colorado State University (CSU). Like its equivalents in other states I passed through, such as Laramie in Wyoming and Missoula in Montana, it had that same college-town vibe. The same charming infusion of intellect, youth and dynamism. An engine of innovation from people of all varieties, pelages and phenotypes. And a lot of watering holes. Thinking is thirsty work.

Fort Collins is actually a very verdant place. It has even recently implemented an urban forestry plan. Called 'Rooted in Community', the plan sets out how the town's more than 500,000 trees will be managed, on both public and private land.[1] There is even a forester to oversee this urban woodland.

I was in the old part of the jungle to meet Mireille Gonzalez. We gathered in one of those numerous watering holes, as the sun approached its zenith and filtered through the emerald canopy outside. There was a particular idea that I wanted to discuss with her: governance. Governance is the management of something – an organisation, a country or even a process – at the highest level.[2] It's a foundation for people with differing perspectives on any given issue to build common ground in pursuit of a common goal. But because it's high-level, it must factor in things like economics, philosophy, ecology and politics as well. Not to mention the historical and psychological dimensions of any social process.

And a quiet corner of a coffee shop in a busy college town was a fitting place to meet for this discussion on governance. No bears – robotic or otherwise – to fend off with mace. No grumpy dogs – livestock guardian, XL Bully or otherwise – to dodge. No bison of any description.

Just conversations about conservation. Concepts to be evaluated. Theories to be ruminated over. Ideas to be refined. As I said, it's thirsty work.

It may not seem like it, but this is the coalface of coexistence. The technical tools that we've encountered, whether Swiss electric fences, Dutch compensation schemes, Wyoming wildlife rangers or Montana wildlife tourism may generate the headlines, and either income or bills. But it is the hard graft of governance that underpins deterrence, finance, force and enterprise. It is the hard graft of governance that provides the frameworks within which managing coexistence occurs. And it is the hard graft of governance that allows the full assemblage of human wildlife, often with more views on large carnivores than you can shake a stick at, to conflict, compete, contest, concede, compromise and, ideally, find consensus of some sort on this contentious topic.

Governance takes place through conversations between people in coffee shops and community centres and fields and boardrooms and lecture halls and pubs and legislative chambers. Often with beverages in hand. It is usually not glamorous. Some might even call it boring. Yet it is absolutely critical for coexisting with carnivores. For, as we've seen on our journey together so far, this coexistence is less about conflict between people and predators, and much more about conflict between people over predators.

So it was fitting that Mireille and I were meeting to have a conversation about governance in Mugs Old Town in Fort Collins. Not only was Ray, as she usually goes by, a director of CSU's Center for Human–Carnivore Coexistence, which works extensively on governance issues in multiple settings. Colorado was also the testing ground for many of the theories and frameworks that we geeked out over. Colorado was home to the most significant current reintroduction project in any of the 50 laboratories of democracy that constitute the United States of America, or, for that matter, the world.

And whereas the famous wolf reintroduction to Yellowstone, which we'll discover more about in the next chapter, had been mandated and managed by the federal government, the unfolding Colorado one was more bottom-up. In fact, we've already considered some of the project's planned innovation around financial compensation in Chapter 7. However, this rewilding experiment on a grand scale was also of interest to me not only because of the predator involved, but also because of the primate involved in orchestrating, resisting and researching it. We are a fascinating species to study. In large part, this is because we don't just have opposable thumbs but also opposable minds. Take any given issue, and any ten humans; throw them in a room and lock the door; and soon you'll have at least eleven opinions. Not just *between* us, but *within* us. No wonder governance of anything can be challenging, never mind when it involves the emotive creatures that are carnivores.

The smell and sound of coffee beans being ground hung in the air as Ray recounted the story so far in Colorado.

'There have been conversations about potential wolf reintroductions to Colorado for twenty-plus years. A working group on the issue was even established in the early 2000s. One of several arguments they made was that public land in the state could and should be used for the conservation of the public trust resource that is wildlife, including wolves. CPW was lobbied about it but did not move forward with the idea at the time.'

CPW was Colorado Parks and Wildlife, the state agency responsible for conservation, hunting and protected areas. I'd visited them for an interview the week before.

'So how did we get from then to now?' I asked.

'In part, this gave rise to the public ballot initiative of November 2020. By fall 2019, almost double the required minimum of 120,000-odd signatures had been gathered.'

Polling put support for the motion at around 80%. In the end, though, the margin was much narrower. With only 50.9% of the three million votes cast in favour of Proposition 114, the poll was even tighter than the UK's Brexit referendum of 2016.

Through her PhD research, Ray had worked with CPW's social scientist to investigate the social outcomes of the resulting 18-month-long stakeholder and public engagement process. This included not only representatives from all sectors with an interest in the reintroduction, but also opinion leaders and the general public. At the beginning, middle and end points of this procedure, the research team explored changes in various social outcomes that, theory suggested, such engagement processes had the ability to effect. These included reducing social conflicts about reintroduction and increasing support for the final wolf restoration and management plan.

Like me, Ray came from a varied background that included conservation, animal welfare and agriculture. But unlike me, she'd also specialised in peace and reconciliation studies. She brought this diversity of subjects to bear on understanding Colorado's wolf reintroduction process, developing a theory of inter-group conflict and identifying the drivers of social conflict.

'Through my own research and that of others, I came to understand that the conflict surrounding wolves is actually symbolic of deeper, identity-based conflicts such as debates about states' rights versus federalism and urban versus rural values.'

From her research, Ray identified three broad groups within Colorado. There was a centrist group able to consider other perspectives. And then there were polarised groups opposed to and in favour of wolf reintroduction. Interestingly, the common ground did not just exist in the moderate middle.

'Despite their differences,' Ray noted, 'the polarised groups have similar perspectives and use similar language when describing the nature of the conflict and perspectives about each other.'

Amidst the background hum of the cafe, our conversation continued. It was relatively easy to sit and theorise from the comfort of our Fort Collins coffee shop. In fact, it was college towns like it and Boulder, as well as the broader Denver metropolitan area, who had voted in favour of Proposition 114. But the Front Range, as this broad valley running the length of the state is known, wasn't where wolves were to be reintroduced. The proposed sites lay in the Western Slope, the less densely populated mountainous parts of Colorado. Here, most voters had rejected the motion. Being from Northern Ireland, I couldn't help but draw parallels with the Brexit vote. It may have been the English shires and cities that voted 'leave', but it was the fragile, fractured communities of Ulster who had to live with its most profound and permanent implications.

How, then, to govern a process like this that was so tightly contested at the ballot box? How to govern a process in which those voting for the reintroduction of a large carnivore species were not the ones living with its return? How to find common ground between the urban liberals of Fort Collins and the rural conservatives of the Rocky Mountains? Was this population of people in Colorado even a single species any longer or was it diverging into two Americas, just as I had grown up between two Irelands, or since Brexit, two Britains? I went to the Western Slope to try to answer these questions.

Hanson country

As I drove west from Fort Collins a few days earlier, the mountain road hugged the river as it sliced through deep valleys. Evergreen forest carpeted the hillsides. Except where the blackened skeletons of dead trees stood sentinel, testament to the awesome power of unchecked forest fires. Yet even here, the first flush of a season of new growth carpeted the understorey. Phoenix-like, life arose from the ashes.

The story of life here is a microcosm of the American West: cattle, coal, railroads. As I crested the Cameron Pass at 10,276 feet and entered Jackson County, the landscape changed. Forest gave way to a vast plain, ringed on all sides by mountains. A geological feature called North Park, it was a kaleidoscope of colour. Granite greys of the peaks, some dusted with the first snow of the season. On their flanks, golden aspen interwoven with dark conifers. Beneath them and stretching to the horizon in all directions, the olive greens of the sagebrush and the earthen palettes of grasses come to the end of the growing season: a tired yellow here, a weary red there.

Before I arrived at my destination of Walden, I stopped on a ridge overlooking Arapaho National Wildlife Refuge. Beneath a blue sky, an equally blue river ran through it, wending its way into horseshoe bends and oxbows. The Native American tribes, including the Arapaho, had called this place 'the bullpen', such was the quantity and availability of buffalo, or American bison. It must have seemed like the herds outnumbered the stars. The story of life here was also a microcosm of the American West in another way: the buffalo were long gone, and with them, the tribes that depended upon them.

In their place had come Hansons, among others. On enquiring about a place to camp in Walden, I was directed to the Sheriff's office. Once I'd registered, I found myself setting up camp in Hanson park. On my way into town, I'd passed Hanson ranch. A walk around the town cemetery revealed a Hanson grave. Amanda Hanson, born in Gotland, Sweden, in 1875, died in Walden in 1902. She was just 27.

Yet again, the past reached out to me. I remembered how profoundly moved I'd been unearthing the seventeenth-century coin from the excavation in Gotland 15 years before. Across the oceans and the centuries, I felt the same again. A connection to this place and to this person. I thought upon the nature of human history across space and time: how tragic and how terrible; how marvellous and meaningful. So often the winner took all. So often there was so little common ground.

This was a harsh and unforgiving place to camp. In the night I awoke, shivering in the half-light of pre-dawn. I'd underestimated how cold it was going to be at this altitude as autumn tightened its grip. My one-season tent and four-season sleeping bag weren't quite up to the task of keeping me warm. In the morning, a layer of ice sheathed my tent and cascaded off the door as I unzipped it. The next night I added a bivvy bag and base-layer and slept soundly through to first light.

This was also a harsh and unforgiving place to make a life. The short growing season limited agricultural productivity to a single crop of hay and a single crop of cattle. All the primary industries were represented here as trucks thundered down main street: farming, logging, mining, hunting. I stumbled into the diner each morning to warm up, cold fingers curled around my coffee cup. The watering hole was packed with hunters here for elk season.

It was the polar opposite of Fort Collins, and not just because of the weather. In place of the skinny oat lattes, plain old drip coffee. In place of the skinny jeans and flip-flops, camo gear and boots. In place of avocado on sourdough toast, chicken-fried steak with biscuit and gravy.

I absolutely loved it. Just as a part of me felt fully at home in the Fort Collins cafe, another part felt fully at home here. At 4,180 km,² Jackson County was going on for a third the size of Northern Ireland. I went looking for one of its 1,379 inhabitants to talk about the wolf reintroduction.

I cold-called a ranch that I'd passed on my way into town the previous afternoon. No one was home so I left a note stuck in the door. I went off to explore the basin and watched the pronghorn skip through the sagebrush. I came back that afternoon. My note was still there because I'd stuck it in the wrong door, but the rancher was home. On condition of anonymity, he agreed to talk to me later that evening.

Technically, as the rancher – let's call him Bob – pointed out, correcting my geography, North Park was not part of Colorado's Western Slope, being just on the cusp of the state's continental divide. The river I'd looked down on in Arapaho flowed to the Atlantic. The rivers just over the mountains flowed to the Pacific. Yet wolves were already here, a small number having moved in from neighbouring Wyoming. Complicating things, this was a reintroduction on top of a recolonisation.

He'd seen them on his land once, but hadn't lost any cattle, unlike a neighbour. And ranchers like Bob feared that more wolves were going to arrive from the adjacent reintroduction sites.

Over beverages at the kitchen table Bob told me his multigenerational ranching story.

Bob wasn't opposed to conservation. In fact, as he pointed out, 'A true rancher is a conservationist: food for livestock provides food for wildlife.'

Nor was he opposed to reintroductions per se. He had fond memories of moose being returned to the area in the late 1970s.

'There was community support and people were excited.'

But predators were a different matter. He'd long had problems with coyotes at calving time. He'd once shot a family of mountain lions that had denned near the ranch, as he'd feared for the safety of his young children.

'There is the same potential risk with wolves and children,' he worried.

I remembered the placards from Switzerland: 'Regulate wolves. Protect children and livestock.' Bob continued, with concerns that I'd heard before in the Swiss mountains and in the Dutch lowlands; in the fields and in the forests; over drinks hot and cold.

'There was a reason our ancestors got rid of the wolf – there was a problem with them.'

'What about these plans to bring them back?' I asked.

'Colorado has too many people and not enough space for wolves,' he argued. 'And it was urbanites who voted for the reintroduction. It doesn't affect their Front Range lifestyle.'

Bob was also concerned that the elk herds, and the elk hunting that was a seasonal mainstay of the local economy, would be negatively affected by the return of wolves: 'In a few years, do you think that restaurant in town will be as full of tourists as it is of hunters today? I don't think so.'

I wondered how the reintroduction planning process had been managed and governed from his perspective.

'The reintroduction process hasn't been managed well. The state is not telling all of the details of the process and it is also being driven by the Governor for personal and political reasons.'

But amidst the frustration and the worry, Bob ended philosophically: 'Ranchers are stewards of the land. If we don't take care of it, it won't take care of us.'

I realised then that the land itself was the common ground I so desperately sought. That governance needed to come from the ground up, literally and figuratively. That nature was the essential strand of human nature, the ultimate arbiter in our long history of disputes, as well as frequently the cause of it. Whether these disputes involved predators or our fellow primates. Whether they took place in an urban jungle or a rural fastness.

It has always been about the land. It is always about the land. It will always be about the land.

The land equalises us all. Princess or pauper, it feeds us, clothes us, inspires and sustains us. We cannot opt out of depending upon it. We cannot opt out of sharing it with others. So let us opt in to treating it well, so that, in turn, the land may take care of us. And let us opt in to sharing it well, so that everyone and everything may flourish on Planet Earth.

In Jackson County's Hanson country it was getting late as I drove back to my freezing tent. I lay awake in it, and not just because of the cold. I felt like I'd answered some of my more theoretical questions but also generated some more practical ones. How to govern a wolf reintroduction process that satisfied Walden and Fort Collins alike? How to bridge the gap between avocado eaters and biscuit munchers? Were there workable governance solutions or was it really an intractable, identity-based conflict?

Back in Fort Collins, I told Ray about Bob. She empathised. Colorado State University itself had been neutral on Proposition 114. As a land grant institution, CSU was empowered and funded to deliver agricultural extension to farmers and ranchers across Colorado, putting research into practice. It also had considerable expertise in carnivore ecology and conservation through the Center for Human–Carnivore Coexistence (CHCC). Not to mention, through Ray and other colleagues, skill sets in subjects, like peace and reconciliation, that were not usually on the reading list for most conservation programmes.

But they should be. They urgently must be. Because the key to saving nature is human nature. And understanding human nature requires a diverse array of subjects and skills that bridge our artificial subject silos in the social process that is conservation. Through the social sciences, history, law, the arts and the humanities, we illuminate the intricacies of our thoughts, words and actions. At the coalface of coexistence, when fused with knowledge of the natural world that comes from the natural sciences, these tools can help us find and govern common ground.

Put another way, we need the natural sciences to help us understand the ground in common ground, and all the human-focused subjects to help us understand the common. And as no one individual, or even organisation, can hold all the relevant skills or experience, we need each other to seek out this elusive place. Even if coexistence is a never-ending journey as much as a destination. A destination that will always recede into the distance, like the endless road across North Park's mountain plain.

At CSU, Ray coordinates the Colorado Wolf Conflict Reduction Group, an informal network of state, federal, academic, NGO and rancher representatives from both the Front Range and the Western Slope. It has several functions, including the establishment of the Wolf Conflict Reduction Fund that we assessed in Chapter 7. It aims to provide rapid deployment of management tools as wolves return to the state, as well as a safe space for a diverse set of stakeholders to collaborate and, Ray admits, sometimes compete. It's small and it's not perfect, but it's a start. It's creating common ground.

Ray concluded: 'In Colorado, we hope we can push the motto of "we can do it better". Our desire, in relation to wolf reintroductions, is to learn from mistakes in other states and places, and to become a model for reintroductions globally.'

Herding big cats

Carnivore coexistence governance geeks are a very rare species. Understandably, therefore, you may not consider governance to be the most riveting topic in this book, or, for that matter, in many other areas of life. But it is in fact one of the most, if not the most, important. Governance doesn't just consider management of something – an organisation, a country or even a process – at the highest level, but also the systems required for this.[3] These systems can include laws and legal frameworks; official guidelines; stakeholder fora; conflict resolution mechanisms; spatial zoning; and others. Understanding what governance is helps us to then consider why it is so important, especially in relation to coexistence between people and predators.

Governing coexistence with large carnivores is complex because these animals do not just wander the landscapes of the world around us. They also wander the landscapes of the world within us. And in our inner landscapes they take on a symbolic quality. We project onto these powerful animals powerful beliefs, values and emotions: we love and we hate; we wonder and we fear; we romanticise and we demonise. And in our inner and outer landscapes alike, our polarised perspectives on predators clash.

No wonder governing coexistence with lions and tigers and bears is tricky. But the fact that governance of anything is a challenge at the best of times, and especially so with species like these, doesn't reduce the need for it. It increases it. In this case, for four main reasons.

Firstly, large carnivores roam large spaces. This often takes them out of the protected areas where even their most ardent detractors usually concede they belong. Over time, it may even take them out of the landscapes that we associate with wild animals, through our historical conceptualisation of the wild as a place that we discussed in Chapter 3. And into places that we don't consider 'wild'. That we maybe even think of as 'domestic'. Think of the wolves in the Netherlands that our farming family encountered in Chapter 7. This means that large carnivore conservation frequently requires the involvement of numerous landowners, often with multiple perspectives on these species. A bit like Bob in North Park. Governance is required to find the right balance between their property rights as private landowners – maybe even livestock owners – and other requirements for conservation.

Secondly, large carnivores are a public good. No one owns wild wolves, lynx or bears. Yet there is often a compelling public argument for their maintenance in, or even return to, landscapes. When public support for this increases over time, such as in Colorado, public representatives listen. Laws are made. Governance happens.

Predators are also an example of the 'commons', things like water and air that typically aren't or can't be covered by property rights – that is, they cannot be owned. And as with many commons over time, notably climate and biodiversity, but also predators themselves, individual incentives can often involve exploiting them as much as possible in the short term, leading to their decline, as Garrett Hardin famously described in 'The Tragedy of the Commons'.[4] His more pessimistic view was balanced by the work of Elinor Ostrom in *Governing the Commons*,[5] in which she described communal systems that have been developed over time to avoid the overexploitation of a resource. Again, with large carnivores as a form of commons, communal systems are required to find the balance between using and losing. A balance called coexistence.

Thirdly, large carnivores cost lots of money. We explored and made the case for market-based carnivore coexistence in the previous chapter. But we also acknowledged that enterprise options have their limits. That is because markets themselves have limits. Very roughly, markets are very good at pricing in financial factors, almost always in the short term and sometimes in the longer term. But they often struggle to, fail to, or imperfectly value non-financial factors, especially social and environmental ones. What price to put on the return of a previously extirpated species, as we have already asked? Or, for that matter, on good governance?

If large carnivores are a public good; if they occur on private land, even if that is contested; if the market alone cannot cover the costs of absorbing their presence, because of the damage they can cause: then there is, I believe, a strong case for public investment in their conservation, and in the coexistence methods required to achieve it. If public money is involved then governance is especially important – though, as we saw in Chapter 7, this can come with its own set of cumbersome restrictions.

Fourthly, large carnivores evoke powerful emotions. As we discussed in Chapter 2, our theory of psychic multiplicity allows us to simultaneously hold a wide range of feelings about these animals. In fact, we are quite capable of loving, hating, wondering, fearing, romanticising and demonising individually. But as we also considered in our chapter on psychology, these emotions are reinforced by others through the stories that shape our worlds and rule our tribes. In Fort Collins, Ray shared the concept of collective emotional orientation with me.[6] People experience emotions on behalf of others in their social group, even if they've never met. The relevant areas of their brain can even light up in the same way as if they'd experienced the issue themselves. For example, a livestock farmer from the Western Slope or rural Ireland may identify strongly with the other farmers we meet in this book. On the other hand, a wolf conservationist from the Home Range or urban Britain is more likely to identify with the carnivore conservationists we also meet.

This helps explain why issues like wolf reintroduction in Colorado become intractable, identity-based conflicts. But in this age of identity politics, we urgently need to find common ground with those who think differently from us. No matter their views on wolves, lynx or bears. No matter whether they're from Walden or Fort Collins. No matter if they're liberal or conservative. We're all human, after all. We all come from the land.

These four reasons – spatial, public interest, financial and psychosocial – may explain at least some of the reasons why governing coexistence with large carnivores is important. In a finite and crowded world, its significance is unquestioned. But how can it work in practice?

As we've already noted, principles and procedures from peace and reconciliation studies can and should be applied to coexistence.[7] A model called Conservation Conflict Transformation, for example, uses a conflict intervention triangle, of 'substance', 'process' and 'relationship', adapted from mediation studies.[8] Another uses participatory planning and workshops with all interested parties to move towards coexistence, recognising that this is an ongoing process affected by limited resources.[9] A survey of over 500 conservationists found that shared governance approaches, which combine local perceptions with technical knowledge, were the most popular option, though there was significant suspicion of political processes, such as ballot initiatives.[10]

What about examples, though, where coexistence has to be created from the ground up as species are reintroduced? Like in Colorado, or, potentially, Britain and Ireland. We've already discussed some of the preparation for governance and management underway with the Colorado wolf reintroduction, including the official state process via CPW, as well as the more informal Wolf Conflict Reduction Group that Ray leads. In our islands, there are useful examples of multi-stakeholder governance involving white-tailed eagles in Scotland[11] and beavers in England.[12] In research on the latter, the authors develop a useful concept they call 'Renewed Coexistence' which acknowledges the additional challenges of governing coexistence with a reintroduced species. They recommend that this process include stakeholder and public engagement; monitoring and research; conflict management protocols; and 'broad perspectives on reintroduction trials'. There are also official codes, or guidelines, in place for reintroductions in Scotland since 2014,[13] and in England since 2021.[14]

From high-level ideals to technical guidelines and everything in between: governance is the foundation of coexistence because both are about sharing time and space, with each other and with predators. As Mark Twain famously quipped, 'Buy land; they're not making it any more.' With a finite amount of space on this increasingly crowded planet, a flourishing future for people and nature alike therefore depends on how we share it. And on an increasingly crowded planet, I'm under no illusions that that is increasingly challenging.

Governance is not all motherhood and apple pie. Issues of carnivore coexistence have a way of getting under our skin and into our psyche. There are competing values and visions – for the future and for the land – that may be irreconcilable. Yet still we must share these places and spaces and times. We must find ways to disagree agreeably.

Governance is not the absence of conflict either. Or of competition. It provides a means to manage both of these in ways that sit within mutually agreed boundaries. Ideally, it leads to some form of compromise or consensus, as elusive as that may be. Even if it feels at times like herding big cats, we must find common ground.

Governance is probably the single most important factor dictating the success of any large carnivore reintroduction in Britain or Ireland. It's definitely not boring. And it means that potentially coexisting with large carnivores again in our islands is less about governing our relationships with our fellow predators, as important as that is, and more about governing our relationships with each other.

This section has brought us on a journey across Western Europe and North America to understand coexistence management and governance in action. To farmers and ranchers and conservationists and rangers and hunters and entrepreneurs. To forests and fields and mountains and plains and villages and cities. Now we travel into the future. We weigh the case for and against the reintroduction of lynx, wolves and bears to Britain and Ireland in terms of ecology, politics, economics and philosophy.

In the future and in the present alike, governance matters. For everyone; for everything; for everywhere.

Whether your natural habitat is urban jungle or rural fastness. Whether your preferred food is biscuit-with-gravy or avocado-on-toast. Whether you drink skinny oat lattes or plain old drip coffee in your watering hole of choice. Whatever your tribe, pelage pattern or phenotype. All of us, without exception, come from, depend upon and one day return to the land. It is the great equaliser. The ultimate common ground.

FUTURE

CHAPTER 11

The Call of the Wild

Three bison looked down from the slight ridge to our right. We were getting uncomfortably close. Like a pilot light coming on, I felt the first prickle of fear deep in my gut.

I kid you not this time. To be precise: these were three *live* bison. To be even more precise: three live *wild* bison. And every step was taking us closer. I was starting to wonder how close was too close. But Joanna was in full flow, animating the finer points of wolf ecology with effusive hand gestures. I assumed she'd seen the bison. She didn't seem to be worried. In that case, bison wrangler extraordinaire that I thought I was, I wasn't going to appear worried either. At least on the outside.

Suddenly, Joanna saw the bison.

'Oh, shoot!' she said. 'We need to get out of here.'

She darted to the left. I darted after her.

'Shoot! Shoot! Shoot!' Joanna was obviously on alert. So was I. Instinct initiated yet again. It opened the valves of my adrenal glands. My surging adrenaline was ignited by the pilot light deep within, the prickly sensation that something wasn't quite right with this situation. Internal combustion happened. The fire of fear coursed through my body. My senses ignited: sound, sight and even smell. I breathed deeper and harder as we sprinted to the treeline. Details were seared into my mind.

Gossamer strands of spider silk strung between grasses. A droplet of dew rolling down a sagebrush leaf. Steam struggling into the heavy air from a pile of dung. The lingering petrichor scent of rain on the soil of a bison wallow.

This was a beautiful day to die.

I glanced over my shoulder as we reached the edge of the clearing. The bison hadn't moved.

'I thought you'd seen them!' I gasped as we recovered our breath.

'Too busy talking!' quipped Joanna as we surveyed the scene before us.

I was with Joanna Lambert, a professor of ecology from the University of Colorado – Boulder, at the epicentre of the world's greatest reintroduction story: the wolves of Yellowstone. And just like an earthquake, its effects radiated out far beyond the national park. In fact, the aftershocks were still being felt right around the world. Even in distant Britain and Ireland, the seismic shift towards rewilding as a vision for the land, and the growing calls for large carnivores like wolves and lynx to restore ecological processes and balance within it, could be traced back, in large part, to this very spot.

And we weren't just anywhere in Yellowstone. We were on our way to see the last remaining holding pen from the 1995 wolf reintroduction, the Rose Creek pen in the Lamar valley. There was only one problem. Between us and the pen was a herd of 50 buffalo, or American bison. Or 100. Or maybe 75. The broad open pasture before us was a maze of undulating hillocks and little ridges. For every bison we could see, there was probably one or two that we couldn't. But I wasn't really focused on counting them. Right now, I was focused on staying alive.

Joanna and I convened a field council on how best to proceed. If at all.

'After coming all this way, I don't want to miss the pen,' I remarked in hushed tones, keeping an eye firmly fixed on the nearest bison. Neither did Joanna. She's not the quitting type.

'And we're so close!' The clump of trees in which the pen sat was only a few hundred metres upstream.

'I think we can do it if we stick close to the treeline,' suggested Joanna.

I nodded.

She continued. 'But we need to be very careful. You see those tails up in the air – those animals are nervous. They're not used to seeing humans in this place. And the rut is just finishing, so they're pumped full of testosterone and more agitated than normal.'

We proceeded with caution. And humility. For arrogance would get you killed in a place like this. There was room for bravery, yes, but not arrogance. Bravery helped calm the raging inferno of fear that blazed within me. Bravery comforted the part of me that wanted to head straight back to the car. Bravery was not the absence of fear but its mastery.

I mastered my fear and we crept forward. Humbly. Our movements were furtive. Our communications were whispered. We melded into the landscape.

Rose Creek burbled downhill beneath us. Thick forest cloaked its banks. Spectral mist haunted its treetops on this damp early morning. It was primeval.

Having an extra foot of height over Joanna, I was nominated to stick my head above the ridgeline and see if we'd safely passed the buffalo herd. We had.

'I think we've passed them,' I whispered as she joined me. We came up out of the steep slope and back onto the more level ground.

This was the second time I'd felt fear that morning since we left the car park at the Yellowstone Institute's Buffalo Ranch. And it wasn't even eight o'clock yet. Only a few hundred metres from the centre, we'd approached a sharp bend in the trail. Rose Creek ran next to it and the sound of running water filled the air.

'This is a classic place for surprising a grizzly,' warned Joanna. 'They won't be able to hear us over the stream. We need to make some noise.'

She shouted and I whistled as we rounded the corner. I put a hand on my bear spray canister for reassurance. Thankfully, no bears were in sight.

Joanna began to tell the tale of the decline of the wolf in the USA. From an estimated one to two million animals in North America when Western European settlers arrived some 400 years ago, the species had declined to about 300 breeding pairs by the mid-1960s in the lower 48 states. Bounties had been paid for varmints like wolves, all the way up from local to federal jurisdictions. By 1926 wolves were gone from the Yellowstone area.

Sneaking through this ghostly landscape, scraping in obeisance to its buffalo kings, I had a momentary flashback. A memory that was not my own. Of how this place must have looked and felt 200 years before. I could feel the power in it. I could also feel its power in the present. And in the future. It was the call of the wild.

As we neared the Rose Creek pen, I asked Joanna: 'To what extent was the reintroduction here an attempt to right a historical wrong?'

'It absolutely was,' she rejoined. 'That was a big part of it.'

As I'd climbed back up the sharp, steep slope from the treeline to the undulating meadow, I had ascended Maslow's famous hierarchy of psychological needs.[1] Feeling safe again, I could afford the luxury of pondering the rights and the wrongs of wolf extirpation and return. The call of the wild was once more a powerful idea, place and process. And in Chapter 14, we will ascend to these lofty metaphysical heights again, as we weigh up the philosophical case for and against large carnivore reintroductions to Britain and Ireland.

But minutes before with the bison, and when calling to ward off grizzlies, I had been right at the base of Maslow's psychological pyramid, concerned mainly with my own survival. Base instinct ruled. The fear was real. I could still feel it in my body, even if my mind had moved on.

Minutes before, this Yellowstone meadow had not been a landscape of inspiration to me. It had been a landscape of fear.

Minutes before, this place had not been a field of dreams where I might find myself answering the call of the wild. It had been a brutal killing field where I might lose my life.

To the best of our knowledge, we are the only species that ascends all the way to the summit of Maslow's hierarchy. From our basic physiological needs that we share with all animal life, through safety, belonging and esteem, the first two of which we share with many higher mammals, all the way to the pinnacle of self-actualisation. But in moments like this, in places like this, the grand accomplishments of our opposable thumbs and our opposable minds are forgotten. We are confronted once more with the brutal laws of nature. We become once more, just like the world around us, red in tooth and claw.

This, in a nutshell, is the science of ecology, in which Joanna was an expert. It is also ecology that is the main rationale for reintroducing large carnivores to my native isles. We delved deeply into the science of life and death. And, above all, the science of fear.

The reintroduction of wolves to Yellowstone is underpinned by two main and interrelated ecological concepts. The first is known as landscapes of fear and the second is called trophic cascades.

The presence and behaviour of the bison herd had altered our behaviour, and, in turn, had determined the parts of the landscape that we felt safe to move through in order to get to the Rose Creek pen. In the same way, landscapes of fear were at work when the presence of wolves in Yellowstone began to change the behaviour of the elk herds, their main prey species.[2] The elk avoided sites with higher risks of being hunted down, especially dense cover and open areas. In turn, this affected the patterns of browsing and grazing in these places. With less pressure from the vast elk herds, the vegetation began to recover. Alongside rivers especially, the recovery of willow firmed up the banks, reducing erosion. Wolves had, apparently, changed rivers. Birds and beavers multiplied. Elk numbers fell. The increased carcasses provided food for bears and ravens and eagles.

A small number of reintroduced wolves – 31 initially, plus their descendants as they reproduced rapidly – had an outsize ecological impact on this ecosystem, as the effects of their presence and behaviour cascaded down through the trophic levels of the food chain to herbivores, other predators, other species and plants beneath. These trophic cascades, combined with a landscape of fear, created a landscape of life.

At least, this is the version of events, popularised by George Monbiot in *Feral*, that has caught hold of the public's imagination. If I had a pound or a euro for every time I told someone what I was writing a book about, and they responded with some version of 'Yellowstone! Wolves change rivers!', my bank balance would be a lot healthier than it actually is. In reality, the situation was more nuanced. And because this idea is so important in the debate about reintroducing wolves especially, but also lynx, to Britain and Ireland, I'd come to Yellowstone with Joanna to understand it for myself.

But as we unpack the intricacies of what unfolded in Yellowstone, and try to apply them closer to home, there is no doubt that when the first wolves were released into this small holding pen in January 1995, something changed. As the call of the wild echoed across the Lamar valley for the first time in decades, a generation of naive elk must have lifted their heads in curiosity at this new and novel sound. How quickly that call came to represent the new landscape of fear they lived in.

'Can you imagine hearing those wolves in this landscape again for the first time?' enthused Joanna as we finished our brief exploration of the crumbling pen, where the animals had been held for a short period of time before being released into this valley.

The wild resurgent

We turned and headed back the way we'd come. By now, though, the herd had spread out even more across the meadow, including to the treeline. Our avenue of approach was now blocked by grazing bison. We gingerly picked our way across the terrain, trying to keep the small hillocks between us and these huge bovines, as I peered over ridges and round corners. If anything, it was even more hair-raising than the previous journey. There was no security of having woodland at our back and a stream to wade through in case we needed to make a quick getaway.

We were surrounded by buffalo on all sides. My senses were in overdrive. My heart pounded in my chest.

Despite several hasty retreats and abrupt changes of direction, we managed to make it to the other side.

'They're really twitchy,' remarked Joanna. 'It's as if they're still exhibiting rutting behaviour even though the rut should have finished.'

As if to illustrate the point, a huge bull began to thrash the sagebrush. A mere hundred metres away, this ton of testosterone raged against the world around him. I could see vegetation flying through the air.

What would that vast head do to a human body? I couldn't help but wonder as we fled this landscape of fear.

Safely past the last of the bison, we reached the trail that headed back down to the Buffalo Ranch. The conversation meandered, like the creek next to us, from ecology to human ecology. We migrated from weighty matters of life and death in the ecosystem around us to weighty matters of life and death in the lives we lived. Perhaps it was a release of pressure thanks to feeling relatively safe again. Bar any grizzlies we might bump into between here and the car.

Somehow we got to talking about ME, or chronic fatigue syndrome (CFS). Joanna had had this crushing condition through much of her late twenties. Myalgic encephalomyelitis, to give it its full title, is a post-viral condition that can be extremely debilitating. Usually contracted after a serious viral infection, the theory is that this disrupts the immune system's normal functioning. A panoply of symptoms can ensue – severe fatigue, sleep disturbances, muscle and joint pain, brain fog, bowel dysfunction; the list goes on, varying from person to person and from time to time. Like long Covid on steroids. Sometimes for years or even decades.

The precise causes of ME/CFS are debated. As are the treatments for it. But what is not debated is that this illness crushes people. It takes their life and wrings it from them. It takes their dreams and disintegrates them. It does not kill physically. But it does kill hope. Slowly and brutally it strangles hope.

The hope of a normal life. The hope of an education and a career. Sometimes, even, the hope that you'll be able to get out of bed and have the strength to walk into the garden to feel the sun on your face.

I know this because my wife Paula developed ME/CFS when she was 16, months before her AS Levels.

'How on earth did you finish your PhD?' I asked Joanna, incredulous.

'I don't know,' she marvelled, 'I really don't know. I just kept going. One day at a time. One foot in front of the other. It was awful.'

Joanna is not the quitting type.

Nor is Paula. For well over a decade, ME/CFS consumed her. For well over a decade, ME/CFS consumed me. Not because I had it. In fact, I was disgustingly healthy throughout. But because I became her primary carer for many of those years, and later for our three children as they were born. What made it so challenging was the complete variability of the condition. On a good week, Paula might be at 75% of normal function and could look after the kids and do housework, just about. On a bad week it could be 25%, and so I looked after the kids and did the housework. And this constant ebb and flow of energy and capacity varied from day to day, week to week and month to month. Sometimes even hour to hour. On top of that, with an impaired immune system, Paula caught every viral and bacterial infection going.

It was brutal.

Years passed. Like trees that bend in the wind, or water that flows round an obstacle, we adapted our entire life around coping with ME/CFS. I chose to do a PhD in large part because I couldn't hold down a regular job and, bar three short fieldtrips to Nepal, could work entirely from home and have total flexibility. We chose to set up Jubilee Farm because I still couldn't hold down a regular job, and, being tied to a farm, could work entirely from home and have total flexibility.

It was brutal.

Time passed. Through the combination of the welfare state – God bless Gordon Brown and his tax credits – and the support of our families and friends, we endured. We are not the quitting types. We just kept going. One day at a time. One foot in front of the other. As the kids began school, Paula began to find the time and energy to begin training part-time as a counsellor. There was no eureka moment of recovery or method of healing. But over time, she recovered to the point where she could complete her training and begin her own small psychotherapy practice, so she could have the flexibility to maintain a delicate work–life–health balance.

I am confident in my ability to express myself in the English language, in both written and, especially, spoken form. But I do not have the words, in any form or in any language, to tell you how difficult it was. The strain was indescribable. Sustained hardship beyond the telling of it.

It was absolutely brutal.

But amidst the harsh outworking of the laws of nature in our lives, a small seed of hope survived. A hope of a life of some kind, if not perhaps a 'normal' one. Buried and dormant deep within, this seed gave birth to a fragile shoot. Amidst the inner landscape of fear – that ME/CFS would never relinquish its iron grip on our lives – it took bravery to tend it. I did so partly by answering the call of the wild whenever I could, slipping out of the crucible and into the fields and the forests and the mountains. I became ruthlessly efficient at getting large amounts of work done in limited amounts of time. We both came to appreciate that health is the most precious form of wealth. Like a tree bending or water flowing, we endured. We were resurgent.

As we've already noted, we humans are the only species that we know of that can ascend to the self-actualisation summit of Maslow's hierarchy of needs – and for all those years Paula and I spent most of our time clinging on for dear life somewhere between the bottom rung of survival and the middle rung of belonging. And with a few notable but very rare exceptions, we also appear to be the only species that takes care of our weak and sick and disabled outside our immediate family groups.

The compassion we extend to our fellow non-familial humans helps to set us apart from the rest of the animal kingdom. The support that Paula and I received from non-family members – from financial support from the welfare state through to, on occasion, access to food banks and, more regularly, help from friends from church and other social networks – combined with help from our families, and our personal resilience and adaptability, saw us through this slow-motion crisis of health, wealth, education and career. In the absence of those things, I don't know how we would have survived. As Paula and I frequently remarked to each other over those years, and as we still do to this day, in another place or another era, we wouldn't have made it. For most of human history, Paula wouldn't have survived the rigours of this life. For us, compassion truncated the brutal laws of nature. Compassion made us resurgent.

Ecology is the brutal laws of nature. Within these harsh and unforgiving constants, large carnivores like wolves perform the function of pruning herbivore populations, helping to maintain a shifting balance of predator, prey and plant. In the animal kingdom, the weak and the sick and the old are precisely the ones that wolves and other predatory species target. Natural selection at work.

Back in the Lamar valley, we drove east from Buffalo Ranch towards Mammoth Hot Springs, where I'd left my car.

'This is the second most productive terrestrial ecosystem on the planet,' shared Joanna, as she gestured at the imposing landscape around us. 'The animal biomass is significant because of the plant biomass, which is productive because of the volcanic soils and glacial till.'

Along the way, we stopped to view the remains of a kill by the Junction Butte wolf pack. Adult and juvenile bald eagles perched on trees, alongside a few ravens, as a coyote tugged at the remains of an elk.

Pointing at the kill, Joanna continued, with an enthusiasm that leavens her formidable store of ecological knowledge: 'This is what's called subsidisation. Wolf kills have contributed to more scavenging grizzlies, eagles and ravens.'

The Lamar river we'd traced, and now the Crystal creek behind the kill site in front of us were the rivers of 'wolves-change-rivers' fame.

'How much of the change here was due to wolves changing elk behaviour and how much to changing elk numbers?' I wondered, voicing a key issue in the debate in Britain and Ireland. A recent *Guardian* column by George Monbiot had made the explicit case for reintroducing wolves and lynx to Britain as a means to control deer numbers, and with it, ecological damage caused by their overpopulation.[3]

'There's no doubt that the elk population has fallen from around 17,000 in 1995 to fewer than 4,000 today,' noted Joanna. 'And an adult wolf needs something like 1.8 elk per month and with an annual average of about 100 adult wolves in the park, that means there are fewer elk and those that remain are more wary of predators.'

With elk fewer in number and harder to catch, Yellowstone wolves were learning how to hunt bison. The behaviour was spreading through wolf culture as individuals moved from the original bison-hunting pack – Mollie's pack – into others.

'But,' she warned, 'it is impossible to disentangle the various population dynamics at work here, be that wolves, drought, vegetation shifts, disease or fire.'

Joanna continued with further nuance: 'Even the beavers that moved in after wolves were reintroduced came partly because they too were reintroduced to the adjacent Gallatin National Forest in the late 1980s to the late 1990s. They were already dispersing into the park via the waterways.'

'So it's a bit more complicated than wolves-change-rivers?' I ventured, as we looked out over one of those very rivers.

She paused. 'It's less wolves-change-rivers and more wolves have been one of several factors in changing rivers. Food webs don't work that monolithically.'

The Disney version, as Joanna calls it – that wolves alone were responsible for the ecological renewal that unfolded here – 'made people think about ecology, and cast a new light on the role of wolves, which is a good thing.'

'But,' she added, 'I'm concerned when science communication doesn't give the most recent and realistic representation of how things are. It plays straight into the hands of anti-wolf people. They can downplay the ecological benefits of reintroductions as romanticised.'

Both Monbiot's *Guardian* column in Britain and an *Irish Examiner* op-ed by Eoghan Daltun,[4] making the case for lynx reintroductions to Ireland, focused heavily on the notion that returning large carnivores would lead to a straightforward fall in the British and Irish deer populations respectively. And both largely dismissed hunting as an effective means to control deer. But the biggest flaw in both arguments seemed to be the inference that deer populations were now surging to catastrophic levels because big predators were absent. Given the timeframes involved – centuries or millennia (and for lynx possibly never even present in Ireland, as discussed in Chapter 3) – I find it a bit of a stretch to equate these two things directly. Other factors are involved. And a lot of nuance.

The idea that reintroducing large carnivores would reproduce the same ecological effects that existed before their extirpation is called reciprocity. A review of reintroduction studies, which also balanced things out by assessing studies where predators were removed, did not find evidence of reciprocity.[5] It noted that the outcomes of both reintroduction and removal were varied and complex. Similarly, another overview of carnivore translocations found that their effects on prey species, and the wider ecosystem, were frequently not assessed.[6] Thirdly, in the case of recolonising wolves in Europe specifically, another meta-analysis, or study of studies, concluded that the evidence for 'landscapes of fear' and similar ecological impacts – or 'non-consumptive effects' – was mixed, context-specific and affected by 'human-related factors' such as hunting and roads.[7]

In short, the Disney version becomes a bit more complicated in the real world. In Yellowstone National Park itself, a number of studies and reviews have questioned the simplicity of the 'wolves-change-rivers' thesis,[8] including interrogating some of the original studies that underpinned it.[9] This debate is a healthy and normal part of the cut and thrust of the knowledge production process, scientific or otherwise. But explicitly – or in the case of Monbiot's and Dalton's articles, implicitly – the 'wolves-change-rivers' thesis is very much alive and kicking in our inner landscapes. Like the call of the wild, the concept has a powerful hold on our imagination. And indeed on the reintroduction debate.

It has become a parable. A myth. Almost a fairy tale, albeit a peer-reviewed rewilding one. So what is the significance of the legend of the wolves of Yellowstone for Britain and Ireland in the twenty-first century?

Landscapes of fear and wonder

As we headed back to Mammoth Hot Springs, I steered the conversation round to exactly this.

'Could wolf reintroductions there work?' I asked Joanna, a Brit who came to America with her parents when she was young and never left, and who is also a leading light in the reintroduction of wolves to Colorado.

'I want the world to be wild. When I go back to England, I find a benign, domesticated landscape that has forgotten what it means to be wild. The landscapes of the UK and Ireland are so utterly transformed that it would be hard to find places.'

Building on the conclusion that both the concepts of trophic cascades and landscapes of fear, which unfolded as they did in Yellowstone, could not simply be cut and pasted to our pair of islands in the North Atlantic, space is the first issue to consider in terms of ecological feasibility. Let's first take Eurasian lynx, a forest specialist. In Ireland, a study found 4,488 km² of suitable habitat patches.[10] In Scotland, the figure was over 20,678 km²,[11] while for the rest of Britain it was 11,369 km², although these two estimates included some overlap in the Scottish Southern Uplands.[12] Numerically, this represents a population estimate of 400 in the Scottish Highlands and of 256 elsewhere in Britain. In Ireland, the potential population was not calculated. However, the various authors concluded, due to the size and fragmented nature of the available areas, that only Scotland, and in the rest of Britain, Kielder and Thetford forests and an area of South East England, would be large enough to ensure viable populations of lynx in the long term.

Secondly let's look at grey wolves, which are habitat generalists. In Ireland, habitat suitability has been estimated at 652 km² in four of its national parks, as well as an unspecified area around them, able to carry an unspecified number of wolves.[13] In Scotland, the habitat projections range between 10,139 km² and 18,857 km² that could support from 50 to 94 packs of four animals.[14] Estimates for England and Wales appear to be lacking. With brown bears, also habitat generalists, no one appears to have modelled space scenarios. Given that the species often overlaps with lynx and grey wolves in Europe and North America,[15] it could be inferred that whatever habitat is available for wolves and lynx could, in theory, also be populated by bears.

As well as habitat, we must also consider food. The space may be available, although there are limited large or interconnected blocks of it. But is the estimated habitat of sufficient quality to provide enough wild food for these animals? This is especially important because in the absence of available wild prey, livestock are more likely to end up on the menu. And as we've seen in our journey together so far, that causes all sorts of problems.

First, Ireland. Regarding lynx, its preferred prey, roe deer, are entirely absent, but there are small and scattered populations of the similarly-sized invasive muntjac deer.[16] Lynx can occasionally take young red deer,[17] but that leaves us with sika, an invasive deer species originally from East Asia. The sika is in between roe and red in size, and there appears to be limited evidence for lynx predation on them – but it is likely that it would adapt to catching them,[18] probably younger, sicker and smaller individuals. Ireland also has fallow deer, in between sika and red in size,[19] and, again, lynx may be able to take smaller or younger individuals. In the absence of definitive data, the best we can say is there is probably sufficient prey for lynx. Wolves are able to kill adults of all of Ireland's deer species and, given overall deer density of 0.62–1.69 deer per km^2 in four of its national parks, there is likely to be sufficient wild prey.[20] No studies have been carried out for bears, which are often more omnivorous than carnivorous. In answer to the question of whether there is sufficient food for them, the best we can say at this point is maybe.

In Britain, there is a broader menu of deer species to choose from for large carnivores. Not only are the roe, sika, red and fallow present but also, in South East England, invasive Chinese water deer, and in south and central England and Wales, invasive muntjac deer.[21] These small-to-medium-sized animals could be ideal lynx prey. It is the densities of these first four deer species that have been used to model the estimated lynx capacity of 400 in the Scottish Highlands, and 50 in its Southern Uplands.[22] Also in much of the Scottish Highlands and Grampians, 'deer densities far exceed the threshold of suitability' for wolves.[23] Estimates for bears also appear to be lacking here. So as with Ireland, the answer with bears is again maybe, but there is probably enough prey available for lynx and wolves in the areas identified as theoretically suitable for their return.

The third ecological aspect to weigh up is the wider benefit to nature. Like the trophic cascades we saw in Yellowstone. As we've already noted, from our visit to the Lamar valley and from scientific studies, reintroducing large carnivores is unlikely to return our ecosystems to their natural state.[24] In part, this is because non-native species have been added to the mix,[25] such as those sika, fallow, Chinese water and muntjac deer. In part, this is also because humans and our activities are omnipresent in these places, or almost so.[26] Overall, as Joanna pointed out, predators are but one factor among many driving the process of dynamic equilibrium in landscapes.

But beneficial ecological changes there will certainly be. Deer populations and behaviour will likely change, for native and invasive species alike. Browsing patterns will almost certainly alter, with potential climate benefits as trees are able to sequester more carbon, for instance.[27] Smaller carnivore species, or meso-predators as they're called, that have become more dominant, especially foxes but also, potentially, badgers, will probably be affected.[28] As a result, there may be more ground-nesting birds and hares, for example. It could be a useful boost to nature, which has been on the back foot in both Britain and Ireland for decades.[29]

But there are a number of caveats that we should acknowledge.

Firstly, it will not be the simplistic Disney version. We should not ascribe to large carnivores an unrealistic ecological saviour status. Wolves will not change rivers in Britain and Ireland. But they may be a factor in doing so as the call of the wild echoes across our islands once more, and some places become again, for deer at least, a landscape of fear. The ecological change may vary from place to place and from time to time. Yet we wouldn't know precisely what was going on for years after, maybe even decades. It would be an ecological experiment of extraordinary proportions. A part of me loves this. As you might have noticed, I have a hearty appetite for risk and can survive, even thrive, on uncertainty. Others will feel the same. Yet others will struggle with the unknown.

Secondly, it will not recreate the past. That world is gone forever, despite the hands of history that keep reaching out and tugging us back. We can only go forward. We can only create the future. And to give rewilding science its dues, it acknowledges this,[30] which is something I really like about it. Increasingly, people want wildlife and wild places in the future they imagine for themselves. Some of their elected officials sometimes listen, which is what we'll be discussing in the next chapter.

Thirdly, the laws of nature are as brutal with reintroduced carnivores as they are with the herbivores they prey on. Failure rates were 17% for small and 25% for large carnivores in one review of translocations.[31]

Reproduction rates – a sign of success – were only 37% in another.[32] Reintroduction failure rates, especially where the ecosystems have changed considerably, have the potential to be higher still. Reintroduced lynx and wolves will probably die in large numbers. At the hands of humans, they will be shot, poisoned, trapped and run over. At the hands of Mother Nature, they will starve, succumb to disease, kill each other and be impaled on deer antlers. Theirs will be a landscape of fear. For them, the call of the wild will be utterly harsh and unforgiving. I accept all of this as part of the reintroduction process. Others will agree. Still others will find this upsetting and even unacceptable, for compassion is as much a part of human nature as our harshness.

The island status of our nations is the fourth and final caveat. As we saw from considering the theory of island biogeography in Chapter 3,[33] the limited spatial nature of our islands increased the likelihood of large carnivore extinctions here and the certainty of them becoming permanent in ecological terms. And it will also make their return more challenging. Their populations cannot be reinforced and added to by natural movement from other countries, especially in isolated pockets, like lynx in South East England. To ensure the genetic diversity essential for healthy populations, repeated supplemental reintroductions may be required over years and decades.

Back in Yellowstone, Joanna and I came round a corner and screeched to a halt. The route was blocked. Not by bison but by humans. Maybe 30 vehicles and 100 or so people. All straining to catch a glimpse of a distant wolf pack. I put my eye to a spotting scope and caught my own glimpse of a magnificent black male resting in the grass, master of all he surveyed. I was filled with wonder. The wild called and my heart answered.

The ecological benefits that wolves like this have brought to Yellowstone have been significant, if nuanced. The ecological benefits that wolves and lynx, and maybe even bears, would bring to the uplands and forests of Britain and Ireland would probably be significant, if nuanced. Despite our dismantling of the simplistic Disney version of events in this chapter, the ecological case for their reintroduction remains strong, if nuanced. Nature is not a quitter; when given half a chance, the wild is resurgent. And everybody loves a comeback kid, though not always a comeback carnivore.

For ecology doesn't take place in a vacuum. It takes place in a political context where perception is reality. What some perceive as tens of thousands of square kilometres of theoretically suitable habitats across our islands, with deer densities to match, others perceive as domesticated landscapes where the wild and its denizens have no place. What some perceive as a landscape of fear for deer only, others perceive as a landscape of fear for them, their children, their pets and their livestock. Especially with wolves, whose howls split the political landscape as much as they do the natural one. Here, the call of the wild is as much a rallying cry against reintroductions as it is for them. Here, the landscape of fear is driven by the brutal and shifting calculus of politics. Here too, political animals roam the landscape in front us, like those bison in the Yellowstone meadow.

Like the wolves of Yellowstone and the echoes of their howls, the wild still calls in our landscapes of fear and wonder. What will our answer be?

CHAPTER 12

Political Animals

The emu hissed. Rearing to its full height, it easily matched my 186 cm. I dodged behind a tree as the bird lashed out with its formidable talons. I backed off further, putting another tree between us. Then I ran.

I ran as if my life depended on it. I ran between the birch trees, bouncing over a tangled web of roots and moss and tussocks underfoot, as the summer sun streamed through the foliage overhead. I ran to the safety of my green static caravan.

I looked back. The emu had not pursued me. I could still make it out, 50 or so metres away, weaving slowly between the trees in its peculiar bobbing gait. As quickly as it had appeared and attacked me, it seemed to have gone back to patrolling the swamp it lived in. I watched it disappear into the woodland. As I recovered my breath and my nerve, I took stock of my surroundings.

I was not in Australia, as you might expect. On the contrary, I was somewhere deep in the heart of our green and pleasant land. It was a glorious summer day, as only Irish or British summer days can be. On the other side of the regular boundary fence, cattle and sheep grazed. Inside the perimeter, exotic wildlife abounded. It was August 2007 and I was working at a monkey sanctuary.

It was turning out to be an interesting summer. I'd been scratched by a bobcat, built cat enclosures, butchered a deer, walked a wolf, hugged a tiger, rescued a swan, been shat on by a very angry snake and fed a caiman, among other things. Health and safety was virtually non-existent. But being 19 and virtually fearless, I was having the time of my life. It was absolutely glorious, as only working with interesting animals, and their equally interesting people, can be.

I'd started the summer with a very well organised and fully risk-assessed placement doing wildlife conservation – highlights of which featured fishing old motorbikes and trolleys out of ponds and conducting dragonfly surveys. I'd then systematically organised several animal husbandry placements to gain experience with a broad spectrum of species. Cats, large and small? Tick. Reptiles? Tick. Odds and ends of everything else? Tick, tick, tick.

Next on my list were primates. I found a sanctuary and arranged a three-week placement. If the other animal husbandry placements were interesting, both in terms of the experience gained with animals and with people, this took the definition of the term to new and uncharted heights.

For starters, when I arrived at the nearest train halt, there was no one to meet me. I had no address or number with me either. Eventually, after a lengthy wait, I was collected. We zipped through narrow country roads in an old army jeep. I remembered the distinctive model and colour from the endless border checkpoints I had experienced as a child growing up at the end of 'the Troubles' in Co. Monaghan in the 1990s.

Next, having arrived late, I wasn't introduced to the pack of dogs that lived at the sanctuary, or they to me. They were all tucked up in bed when I splashed through the swamp to my caravan late on. The first thing I knew about them was the next morning. As I walked over from the static to the main house, a ferocious baying erupted. Eight dogs came charging towards me. It took all my willpower and knowledge of dog behaviour to stand my ground. I let them sniff me and we became instant friends. The only nips were playful ones from the Japanese Akita, still a puppy in an enormous dog's body.

Then there were the emus. There were three of them. Two of them were very friendly. They'd even wake me in the morning by tapping on my caravan window. I could hear them coming before I saw them because they all made this very distinctive sound, like a basketball being slowly bounced inside a barrel. But one of them was the demon emu that we've already met. When its mate died, it became very aggressive towards people. The problem was that it looked and sounded just like the friendly emus. Until it came flying at me, trying to split me open from sternum to scrotum with its pointed talon-toe, delivered on the end of a flying roundhouse kick. The other problem was that I didn't find out about this emu's vindictive nature until after its first attack.

If you think this is weird so far, prepare yourself. I'm only getting started. It got even weirder.

In the house, amidst piles of boxes, a bevy of free-flying parrots screeched at each other. Here lived the first of the sanctuary's primates. As I sat on the sofa, the De Brazza's guenon sat at my feet and groomed my leg hair. Then my arm hair. Then my head hair. My hair has never looked so good or been so clean. Like the mini Confucian master it resembled, this mid-sized African monkey and former pet ruled over the house and all its inhabitants.

Near the house were the only primates at the sanctuary that lived in an enclosure. A pair of Japanese macaques formerly from a circus, these large Asian monkeys were the bully boys around these parts. Naturally, I was handed a bucket and shovel and told to clean them out. The really aggressive one was locked inside, leaving me with the less aggressive one sitting at the top of the cage looking down on me as I shovelled monkey faeces beneath him. Of all the captive animal experiences I had that summer, the next summer or since, this was my most uncomfortable. I felt extremely uneasy. Maybe because the threat was three-dimensional rather than two-: the entire time I kept waiting for a mad macaque to drop on me from above, with a flash of yellowed canines and a simian screech. I shovelled and scraped as quickly as I could, and got out of there as quickly as I could.

Next, we were off to feed the sanctuary's main group of monkeys. The ex-laboratory capuchins lived on forested islands in the middle of a small lake that had been excavated for this purpose. A spring kept the lake full, and the surrounding birch forest damp, at all times. These South and Central American primates climbed trees, stole waterfowl eggs and generally hung out to their heart's content, with small huts for shelter. As we poled out to the islands in a small rowing boat to give them their breakfast, the health and safety briefing was reassuringly short: 'If the monkeys attack, jump in the lake.' It was a good thing I could swim.

Feeding all these Old and New World primates, plus emus, dogs, poultry and several horses, not to mention the humans, was no mean feat. Thankfully, the sanctuary had an arrangement with a local supermarket. Several times a week, we visited at closing time and raided the skips at the rear of the premises. Into the back of the jeep went fruit, vegetables, eggs and seeds for the monkeys. Also into the back of the jeep went whatever looked suitable for the humans: breads, ready-meals, tinned foods of all descriptions. On one visit we even took home a haul of sirloin steaks, destined for dumping but still perfectly good. I have rarely eaten so well. Human and non-human primates alike feasted on all this abundance.

The fact was, thanks to the temptation to take all this free food, there was actually too much lying around. That explained some of the boxes in the house. It also explained a shed full of rotting fruit that I had to sort through. Amidst green clouds of fungal spores, I sifted through what was still fit for feeding to the monkeys and what was only fit for dumping out the back. I managed to find a mask for some of the several days I spent doing this. I coughed and spluttered. New species of goodness-knows-what probably evolved in my lungs. It may have taken years off my life.

Then there was the caravan. The detritus and secretions of previous generations of volunteers littered and stained it. It made the student digs I was used to look like a five-star hotel by comparison. I shudder to think what new species were definitely evolving under the bed and in the sink. The bathroom could have kept an entire team of microbiologists busy for a month. I gingerly cleared a space on the bed for my sleeping bag and excavated a pathway to and from the door.

The friendly emus kept knocking on my window at the crack of dawn. The unfriendly emu kept trying to kill me most days. I became very good at sprinting over uneven ground without spraining my ankle. My hair was scrupulously free of parasites. My stomach was full of steak.

It was a truly bizarre experience in a truly bizarre place.

I learned a little about nature. I learned a lot about emus and monkeys. I learned even more about human nature.

After two weeks I'd had enough. I who rarely quit, quit. A friend had belatedly asked me to play bass guitar at his wedding in Belfast and I used that as my excuse to leave a week early.

The sun still streamed through the birch trees as I bid farewell to the swamp and all its inhabitants.

Exactly a year later, I went to work at a Canadian zoo from July to September. I had to look after emus again on one of the three hoofstock sections that I worked, as well as some ostriches. This time, though, forewarned was forearmed. And this time, the facility, although also privately owned and operated, was much better run. As part of the health and safety approach, all zookeepers had to carry a broom or rake with them when entering a pen with animals. The idea was that the long-handled implement kept a potential attacker at arm's length. Better that an inanimate object took the force of a headbutt, kick or bite than your arm, leg or torso. The one time I went into an enclosure without a broom, to close the door to the scimitar-horned oryx stable so that I could clean it out, I got a dressing-down from one of the senior keepers. Although I protested that I was just trying to save a bit of time in what were long and busy days, and that all the oryx herd were outside, his retort made sense when I thought about it:

'You don't want a pair of those horns going straight through you, do you?'

I always carried a long-handled implement after that.

Some animals were still too dangerous to go in with, even with a broom or a rake. Like the ostriches. Or the rhino or Cape buffalo that we met fleetingly in Chapter 3. In theory I should have carried one when I opened the gate with the first set of bison we encountered, also in Chapter 3. But there was no avoiding the emus.

I opened the enclosure door and stepped inside with their food. Sure enough, one of the emus came at me: raised to full height, hissing, kicking. But I was waiting for it. I dropped the bucket of food and jabbed towards it with the brush held firmly in both hands. The broad head caught the emu smack in the middle of its chest. Down it went. On its back. And it lay there, legs flailing in the air, unable to get up for a number of seconds.

I may have said something very rude to it. And maybe something about karma.

Either way, I had faced down my emu nemesis. They didn't bother me again after that, though I always had my broom to hand, just in case. In turn, I had no need to shout at or be rude to them.

I just spoke softly and carried a big stick.

The law of the jungle

The political phrase 'speak softly and carry a big stick' was popularised by Theodore Roosevelt at the start of the twentieth century.[1] Appropriately for our story, with its African undercurrents of both my individual upbringing and our collective evolutionary past, the expression was originally a West African proverb, as Roosevelt pointed out when he first used it. The saying refers to the prioritisation of diplomacy in international affairs, alongside the capacity to use force as a last resort.[2]

It is a pragmatic political philosophy for a pragmatist like myself.

I grew up in a very conservative household. Yet my dad, a Presbyterian minister, preached compassion for animals and nature from the pulpit long before it was fashionable. He even brought a live terrapin into church one Sunday to illustrate a sermon point. In my twenties, and, ironically, while doing very well for myself at business school, I turned leftward. But despite all my left-wing zeal and my critiques of capitalism, I never lost my childhood interest in enterprise. I settled on social enterprise as a marriage of entrepreneurial dynamism with civil society's values and ethics. Jubilee Farm was the result.

But during my three years of creating the parent organisation, and especially during my five years of establishing and running the farm, one thing changed. One thing was radically altered. One thing made me tack rightward again, ending up as the pragmatic centrist that I am today. That one thing was my experience of dealing with government.

Throughout my twenties, I had believed in and argued for the transformative power of government, and, in particular, of its central role in regulating the market to account for its failures and shortcomings. I believed in regulation. I still believe in appropriate regulation where necessary. But my experience of being regulated was not a happy one. In fact, in experiencing regulation by government in various forms while setting up Northern Ireland's first community-owned farm, something died inside me. My naive belief in government as a predominantly good thing was extinguished.

It wasn't just the endless and pointless farm paperwork that filled my days. And I write that as someone who actually likes paperwork, generally speaking.

It wasn't just the endless round-robin phone calls with agricultural bureaucrats, like the horror story in Chapter 1, as seemingly straightforward requests were transformed into mind-numbing slogs through the bowels of administrative hell.

It wasn't just the fact that to erect a twenty-foot by eight-foot lean-to greenhouse against the farmhouse gable wall, we had to employ an architect to get it through building control, because they wouldn't accept my attempt at filling in the relevant form. Or the fact that we then had to employ a chartered consultant engineer to certify that the greenhouse wouldn't collapse because we'd neglected to show them the foundation – all paving slabs and reinforced concrete – before we built the frame. You could have driven a Challenger 3 tank over that foundation and it wouldn't have cracked.

It wasn't just that when we tried to apply for planning permission to build a barn, not to expand our livestock numbers but simply to overwinter the small goat herd and fattening pigs that we already had, instead of relying on polytunnels, we had to wade through a statistical model for ammonia emissions that made several people with PhDs throw up their hands in despair. Or that as a small social enterprise operating on a shoestring budget, we had to then fork out for a specialist emissions consultant, over and above an architect and a planning consultant, to do battle with another statistical model entirely. And all these rules, designed to minimise the effects of large agro-industrial farms on adjacent biodiversity, applied with utter inflexibility to a tiny agro-ecological farm busy *expanding* wildlife habitat *outside* the adjacent nature reserve.

It wasn't just all that. More than that, it was the underlying attitude. An attitude that I encountered again and again, over and over. An attitude that broke my belief in government. The attitude was: 'The answer's no. Now what's the question?' It is not an attitude to build a better world by. It is not an attitude to tackle the greatest challenge of all time, namely reconciling ecology and economy so that they are synchronised instead of segregated. It is not an attitude with which to get stuff done.

It is a miserable, simpering, ugly attitude that is the antithesis of innovation and entrepreneurship. And I cannot stand it.

To be fair, this attitude specifically, and bureaucracy in general, is not just limited to government. On changing bank accounts for Jubilee Farm on one occasion, given our legal form was an unusual one called a Community Benefit Society, I had to inform a business banking manager at a major UK banking conglomerate that the Northern Irish bank to which I was attempting to switch our account was actually a fully owned subsidiary of theirs and not a separate firm entirely. Civil society organisations can also be bureaucratic in their own way. As can the dreaded NIMBYs, mentioned earlier, who just don't want anything new built anywhere ever.

Nor is innovation impossible for governments. Far from just being the lenders, menders, spenders and defenders of last resort, they can have a proactive role in leveraging investment from the private sector and civil society, in de-risking early-stage innovation, and in funding public goods, like wildlife and climate and health. During the chaos of the Covid-19 pandemic, we benefited from several rounds of government support and stimulus at Jubilee Farm, though it usually worked best when it was administered by a civil society or other arm's-length entity and not the government itself.

Government, in all its many forms, can even be full of good and well-intentioned people. There are even many fine individual bureaucrats. I know some of them. I met some more on my fieldtrips for this book. But collectively and institutionally, their good characters and intentions are often overridden by and subsumed within the dreaded bureaucratic blob.

Admittedly, then, government does not have a total monopoly on this attitude. But it does, in my view, have a significant and controlling majority shareholding in it. The reasons are many: fear of failure; organisational culture; lack of accountability; limited vision; the wrong mix of carrots and sticks. But whatever the reason, the effect is the same. This asinine attitude stifles innovation. And if there is one thing we need more of in order to tackle the challenges of the twenty-first century, one thing we can all agree on whatever our political preferences, it is innovation.

I mention all this in this chapter on reintroduction politics for a number of important reasons.

Firstly, to be open and transparent about my political views, biases and journey. As it relates to large carnivore reintroductions in Britain and Ireland, I am sceptical of the role of government. Paradoxically, I also know that they can't happen without government.

Secondly, to make the incredibly significant point that no political party, philosophy or position has a monopoly on nature and its conservation. The left do not own virtue, nor the right morality. The centre, smug in its ability to pick and choose from the political smorgasbord, cannot monopolise either. All the great political traditions – conservatism, socialism, liberalism, anarchism and green – find differing ways to value nature, based on differing values. Sometimes these compete. Sometimes they clash. Sometimes they even coalesce. Whether I was right-wing, left-wing or centrist, my own deep love for and devotion to the natural world has never wavered. Whether I engaged with communities of geography or interest that reflected these political positions, or transcended them, I found good people who felt the same about the world around us. In our increasingly polarised polities, this is important common ground to stand firm on with resolve, as we considered in our chapter on reintroduction governance.

Thirdly, politics is in a state of flux. Politics that were once considered left-wing are now touted by right-wing politicians. And vice versa. The political tectonic plates are shifting beneath us, not just across our islands but across the world. Political earthquakes and aftershocks are likely to continue. Ecological processes – including the return of big predators, naturally or with human assistance – do not take place in a political vacuum, but in this unstable political context that is rapidly changing. And there is an increasing risk that these environmental processes, or specifically political action to safeguard them, are seen as left-of-centre issues, when in fact they underpin all human life and activity.

Fourth, despite all that I have written, the role of government in re-introductions is key. Somebody has to be the neutral arbiter of this controversial idea and its potential implementation, as well as of its often combative proponents and detractors. If it is to be implemented over large amounts of land and over many decades, the resources of the state are probably going to be required to make it work well, and to successfully fund the suite of management and governance tools that we explored in section two of our journey.

Fifth, the politics of reintroductions are very, very complex. It makes the relatively simple ecological science of trophic cascades and landscapes of fear seem like a children's nursery rhyme by comparison. Add to this the heady mix of competing nationalisms that have defined the politics of our island nations for the last quarter of a millennia, and beyond. We'll consider these in greater detail later in the chapter but for now, suffice to say they complicate things even further.

Sixth, as reflected in the 'speak softly and carry a big stick' phrase, politics is a curious mix of lofty idealism and unsentimental realism; the law of the jungle holds sway, underpinned by the constants of human nature and the laws of nature. As elsewhere, there is also the same inherent tension between individualism and collectivism that we discussed in our chapter on the use of force. As Rudyard Kipling pointed out in his poem called 'The Law of the Jungle', 'the strength of the pack is the wolf, and the strength of the wolf is the pack'. Overall, therefore, there are a lot of 'isms' to factor into the debate at hand. Furthermore, to go back to our psychic multiplicity model from Chapter 2, we might even harbour competing beliefs on all these political elements within, never mind between, ourselves, as we may also do about large carnivore reintroductions to Britain and Ireland in general. In short, politics is complicated because we, as humans, are complicated.

Lastly, all ecology is inherently political. That is because living among it and using it to survive and thrive, as we must, involves basic questions of who gets to do what with it and when. In other words, control. Or power. And all ecology is also inherently economic. That is because, as assessed in our enterprise chapter, all human economic activity mirrors, or should mirror, nature's economic activity: ecology. In other words: inputs + processes = outputs. As the US politician and founder of Earth Day Gaylord Nelson famously quipped, 'The economy is a wholly owned subsidiary of the environment, not the other way around.' This brings us into the realm of political ecology, which underpins the remainder of this chapter.

Political ecology is a diverse approach to understanding how power and economics shape and cause environmental change.[3] It's a critical approach, meaning it tends to interrogate the status quo, particularly capitalism. But whether you're a fan or a critic of this particular 'ism', or have mixed feelings about it, political ecology is still a useful tool to understand how ecology, including that of wolves, lynx and bears, is shaped by non-ecological human forces. Remember the PPE+PE acronym from Chapter 2 as a handy reminder of the five factors – politics, philosophy, economics, psychology and ecology – shaping large carnivore reintroductions to Britain and Ireland? Political ecology directly underpins two of these, and we'll consider the economic angle in greater detail in the next chapter.

One historical example of political ecology from our story so far is the Cromwellian campaign to exterminate wolves from Ireland, mentioned in Chapter 3. This occurred less because Oliver had read too many fairy tales as a child, and more because removing wolves accelerated the symbolic and actual colonisation of Ireland.[4] Another good example of a political ecology perspective that we've already employed without using the term can be found in the stories of Sarah in Switzerland, the van de Weterings in the Netherlands and Shaun in Wyoming. Each of these cases, although the amount of land and livestock varied considerably between them, were all characterised by a significant degree of suspicion, or even hostility, to the return of the wolf. Yet on closer examination, these feelings were less to do with the threat that wolves and other predatory species pose to their livestock, especially sheep, and more to do with a range of changing political and economic conditions, in the face of which they feel fairly powerless.

They all farm marginal land in marginal parts of their respective nations, where profit margins are tight at the best of times. The adaptations to these delicately balanced ways of working and living imposed by the return of wolves, mediated through the forces of usually urban and remote government and civil society, are disruptive. To them, wolves symbolise a much broader range of underlying changes that are also and simultaneously disrupting their livelihoods, from expanding environmental regulations and rural depopulation to trade deals and subsidy reform.

And none of them liked bureaucracy. On this count alone, we got on famously.

Viewing the issue of large carnivore reintroductions through the lens of political ecology helps bring nuance to the complex set of variables at work in this debate. And because all these things have been building up over time, it also helps to add a historical element to the analysis too. Past, present and future, political ecology delves behind the headlines. It goes behind the scenes. It lifts the lid.

What it doesn't do is generate simplistic social-media-friendly headlines to hurl at opposing echo chambers in our increasingly polarised political debates. Which is precisely why I think it's so important here.

And of all the aspects of ecology, few are more political than large carnivores. They are the original political animals.

The decisions about their potential return to the Republic of Ireland and to the United Kingdom of Great Britain and Northern Ireland will be shaped by the political animals of Dublin, London, Belfast, Cardiff and Edinburgh. In turn, their decisions will be shaped by the laws of the jungle as much as by the laws of the land, that ruthless political calculus 'as old and as true as the sky', in Kipling's words again. But because both nations are fortunate to be democracies, their decisions will also be shaped by social trends, as the elected respond to the electorate. In the remainder of this chapter we will therefore consider the various social and policy trends shaping the large carnivore reintroduction debate, and weigh up the related political cases for and against.

A window on the future

The first trend to consider is the levels of public support for reintroductions. A 2020 YouGov poll of 2,083 adults in Britain found that 36% of respondents wanted wolves and lynx to return, but only 24% wanted bears back.[5] The level of support for wolf reintroduction was also found to be 36% in a much earlier Scottish survey in the 1990s.[6] However, given that lynx have been the subject of recent reintroduction proposals in both Scotland and England, there is more data available for this species. An independent poll commissioned by the Scottish Rewilding Alliance in 2021 found that 52% of Scots supported a trial lynx reintroduction, while 19% were opposed.[7] A representative sample – meaning a more balanced cross-section of society – from a study of 1,042 UK participants found 49% supported a trial lynx reintroduction, 21% disagreed and 30% were neutral.[8] There appears to be no polling from Ireland so far for any of our three species.

Other studies have delved into the reasoning behind these opinions. Almost 80% of 130 participants in another journal article cited 'risks to farming activities' as the primary concern over reintroducing lynx.[9] In terms of benefits, tourism, ecological effects and deer control were listed as the top three. More recent work by the Lynx to Scotland project has added qualitative depth to this quantitative breadth.[10] Their research grouped opinions into five distinct sets: those who supported lynx reintroductions for their potential ecological benefits; those who were supportive for the potential tourism benefits; those who were strongly opposed; those who welcomed the dialogue but perceived significant barriers to the proposal; and those who believed that the case for biodiversity benefit from lynx reintroductions was not strong enough, but who were willing to continue the conversation.

Rural and agricultural communities appear to be less positive about reintroductions. In the 1990s Scottish study, the rate of support fell to 17% among residents of the proposed reintroduction site.[11] In a non-representative sample of 9,600 UK participants, 39% of those from an agricultural background supported lynx reintroductions, while 60% were opposed.[12] A recent sample of 78 individuals from across a broad spectrum of society found that 70% of farmers strongly disagreed with lynx reintroductions.[13] In another similar study, many farmers suggested that they would cull lynx if they were reintroduced.[14]

My own dialogue on this issue with the five main farming unions of Britain and Ireland, plus some livestock and rewilding organisations, adds some nuance to the headline figures.[15] Concerns raised by the agricultural organisations included: that our landscapes have changed and are no longer suitable for large carnivores; that legal protections would limit the ability of farmers to control problem animals; that tourism revenue projections were unrealistic and unlikely to benefit those landowners most affected; and that compensation schemes would be bureaucratic, expensive and financially unsustainable. In putting these concerns to the rewilding organisations, there was considerable appreciation for these positions, including an acceptance that lethal control would be required as a last resort for problem predators. The rewilding organisations focused on the ecological, tourism, deer control and philosophical arguments for reintroductions.

So far, so similar to the three main countries I travelled to in order to research this report and this book: Switzerland, the Netherlands and the USA. We can also look to places like these for hints about how attitudes might vary over time, given that we've made the case that reintroductions in Britain and Ireland may occur over years or decades to ensure a genetically healthy predator population. An overview of 105 surveys in 24 European countries found that attitudes to bears improved over time, while attitudes to wolves became less positive the longer people coexisted with them.[16] The article did not consider attitudes to lynx. From a 2017 study in Switzerland, prior to the recent rapid expansion of the country's wolf population, we can see the same gap between habitat suitability and local support that also likely exists in Britain and Ireland.[17] In other words, there was limited common ground between places where wolves were likely to return and where locals supported their return: a 6% overlap between 'areas of suitable habitat' and 'areas where the wolf was accepted'. It's possible that studies like this could be a window on the future of post-reintroduction Britain or Ireland. In all probability, just like pre-reintroduction Britain and Ireland, perspectives will vary. Those most affected and living closest to reintroductions are likely to continue to be the most opposed, though, again, it's hard to predict things precisely.

From the limited research and polling conducted so far in Britain, support for large carnivore reintroductions in our islands appears to be roughly balanced between those who favour them, those who are opposed and those who are neutral. There is also a probable gap between rural and urban areas. Attitudes in Ireland remain unknown though may be similar to British perspectives. Though difficult to predict, the levels of support may increase over time in both jurisdictions, especially as parallel environmental concerns about nature and climate gather momentum. Directly and indirectly, these may create the conditions, particularly reafforestation of upland areas, that would facilitate large carnivore reintroductions in the coming decades. This is something we'll consider in the next chapter. In the short term, though, we next need to consider the policy and legal context for potential reintroductions.

Ian Convery and colleagues set out a preliminary overview of legal considerations for wolf reintroductions in the UK, but these principles could also be applied to lynx or bears.[18] They also point out that species reintroductions are matters for the devolved administrations within the UK, in which nature conservation laws can vary according to the jurisdiction. However, their assessment is a useful first step in considering the various legal dimensions overall. Convery et al. suggest that, under the terms of the 1979 Bern Convention, transposed into the EU's 1992 Habitats Directive, and then into UK legislation in various forms post-Brexit, wolves would not automatically be a protected species. However, they argue that the Environment Act of 2021 could be used to make a case for reintroductions as part of a broader move towards nature recovery. Thirdly, they point out that licences from the relevant environmental authorities in each of the UK's four nations would be required to proceed. Appropriately for our chapter on politics, they caution that this decision 'may ultimately be political, in the sense that in England, for example, the Secretary of State is able to take the decision out of the hands of Natural England'. Lastly, Convery and colleagues note that various pieces of minor technical legislation, such as the 1976 Dangerous Wild Animals Act, may need to be amended to facilitate reintroductions.

In Ireland, in the apparent absence of any legal review or opinions, we could surmise that the legal situation would probably be similar, given the common underpinning of European law. For example, where

England has its 25-year 'A Green Future' environmental plan, which does mention species reintroductions,[19] the Republic of Ireland has its 4th National Biodiversity Action Plan, objectives of which include 'Meet Urgent Conservation and Restoration Needs' and 'Strengthen Ireland's Contribution to International Biodiversity Initiatives'.[20]

Overall, therefore, we can see that while large carnivore reintroductions may be technically feasible in legal terms, there are already a number of areas that stand out as potential political flashpoints in both Britain and Ireland. First, reintroduced predators may not automatically be afforded protected status. Expect an almighty row about this. Second, as we discussed elsewhere in this book, the extent to which large carnivores will bring about or are required for the desired and mandated nature recovery is debated. Expect more rows about this. Third, there is considerable scope for the politicisation of the process, especially if reintroductions are required to take place repeatedly over time, and administrations, and their priorities, change in the intervening years or decades. Expect fireworks. Fourth, large carnivores can and will cover large distances, including potentially across the patchwork of political jurisdictions that make up the islands of Britain and Ireland, where permissions and licences for reintroduction may vary. Unholy rows are guaranteed.

This last point is worth dwelling on further. The islands of Britain and Ireland contain no less than five major, as well as several minor, political jurisdictions, and two sovereign nations, all with their own unique historical and political contexts. Brexit has stirred up this hornet's nest of political entities. Add the usual combination of parties across the political spectrum, and it all adds up to a lot of politics. And a lot of political animals stalking the landscape, many of them opportunists unmoored from ethics or principles other than the pursuit of power for their own ends. Emus abound. Frankly, it makes the monkey sanctuary-swamp that I lived in for two weeks look like a paragon of reason, order and professionalism.

In this window on the future of large carnivore reintroduction politics, there will be a constant tension between idealism and realism. The laws of the jungle will drive hard bargains as political pacts are made and broken. Trial lynx reintroductions especially will have the potential to be used as a bargaining chip where green parties act as political kingmakers, particularly in Scotland and the Republic of Ireland. If you think this is fanciful, the polling we discussed earlier in this chapter by the Scottish Rewilding Alliance was part of their call 'to make the trial reintroduction of lynx and the widespread relocation of beavers a core part of any agreement they [the Scottish Green Party] reach with the Scottish National Party' in 2021.[21]

At the same time and as we've already hinted at, growing environmental consciousness, especially among younger voters, may drive increasing levels of support for reintroductions over time, in parallel with action on habitat restoration and afforestation in upland areas. This may shift the Overton window – what is considered politically acceptable to the majority of the population – in relation to this issue. Yet set against this will be the political ecology of smaller-scale and marginal farmers in these same upland areas, and the shifting socio-economic pressures that they continue to face as subsidy support changes. In the next chapter we'll explore the economic aspects of all this in detail.

The ecology of Britain and Ireland is a political ecology.

The potential reintroduction of wolves, lynx and bears is a political ecology, just as these predators are political animals.

For these reasons, I don't think that the political case for reintroductions is a particularly strong one, at least not at the moment. In fact, I think that this debate will drive a lot of bruising political conflict across the complex political ecosystems of Britain and Ireland.

Yet I'm idealistic enough to believe that speaking softly around the politics of reintroduction is the best way forward. But I'm realistic enough to believe that most political animals involved will also carry a big stick in the debates, discussions and negotiations to come.

And wherever we are on the spectrum of politics, we should all watch out for emus.

CHAPTER 13

Home Sweet Home

The avalanche thundered. Down the flank of the great white wall, hundreds of tons of snow and ice cascaded. Underlying rock and scree were violently abraded. As it continued its onslaught down the vast north face of Annapurna III, this deadly dervish swept all before it. Even as it reached the terminus of its charge, it ripped up battalions of birch trees and flung them at its feet. Nothing could stand before it.

This was power. Raw and extreme power. Power of Himalayan proportions.

Like a chandelier smashing on a ballroom floor, the extinguishing of these avalanche energies sent towering plumes of snow and dust into the still mountain air. With it came the noise. As we watched from across the Upper Marshyangdi valley, a symphony of sound hit us like a wave upon a shore. A low rumble, like distant thunder. A low and palpable roar, like the occasional earthquakes that shook the windows of my bedroom in Malawi. It shook us out of our monotony. The sound of silence that lay heavy upon this Himalayan heaven was split asunder by the sound of a mountain dying.

Suddenly, it was quiet once more. A temporary truce was called between the forces of geology that continue to push the Himalaya upwards as the Indian subcontinent slides under Asia, and the forces of glaciology that try to tear them down. An armistice of Himalayan proportions.

In the ceasefire that followed, the silence was the sound of a mountain living. In the great valley at its feet spring continued to be sprung from its winter prison. It ricocheted back in riots of green after a season of faded splendour. As the flush of plant growth blushed upwards across the hillsides, the blue sheep followed. With them came the snow leopards. Followed by the herders with their sheep and goats and yaks. And then the tourists, on their circuit-pilgrimage of the mighty Annapurna massif. The intimate and ancient balance between the natural economy – ecology – and the human economy unfolded around us.

It was a scene of truly Himalayan proportions.

I first visited Nepal's Annapurna Conservation Area (ACA) in autumn 2013. Alongside Sagarmatha – or Everest – National Park (SNP), it was one of my two PhD field sites. In my initial visit, I came to scope out my study's logistics and questions through a series of meetings, pilots and interviews. While I was there, and inspired by something I'd read about 'predator-friendly' certification (remember Chapter 9?),[1] I thought up a conservation incentive scheme to tackle the impacts of livestock losses due to snow leopards on the resident communities. By certifying local livestock products (meat, dairy products and fabrics) as 'snow leopard-friendly' and marketing them to the 50,000-plus tourists who pass through the area every year, money could be raised to a) increase the income of local herders, including through increased compensation for losses; b) contribute to community-led conservation and development activities; and c) educate tourists about snow leopards and efforts to conserve them.

Realising that market research was needed before the idea went any further, I added a section to my PhD study's household questionnaire to gather opinions from locals – the supply side of the equation – on the proposed scheme. It was much like the section on the proposed blue sheep translocation that I added to the otherwise identical SNP survey (remember Chapter 4?). I'm a great believer in the pursuit of knowledge for its own sake: if I had a parallel life I'd be a full-time medievalist in it. But I'm also a great believer in applied research that helps to solve real-world problems.

My high-school friend and fieldwork manager Maurice Schutgens led on developing a survey for tourists – the demand side of things – and also added a section on tourists' willingness to pay a theoretical snow leopard conservation premium or surcharge, on top of their existing

ACA entry fee of around US$27. Finally, we added in some questions to gauge tourists' knowledge of snow leopards so that we had a baseline against which to compare the effect of any future conservation education programmes. Two Nepali colleagues, Som Ale of the University of Illinois and Nabin Baral of the University of Washington, helped us to refine the survey by adding parts of a similar questionnaire they'd previously used in the Everest region.

So far so good. Then Maurice had a methodological light-bulb moment. Motivated by not wanting to lug several hundred tourist surveys around Annapurna, in addition to the 450 for local households that we were already carrying, he devised a laminated version of the questionnaire that we could fill in with permanent marker and then wipe clean. Not only did it save on weight, it also saved on paper. As we needed only 20 copies instead of 450, several rainforests could be left intact, which is always a good conservation outcome. A quick pilot later in Kathmandu and we were good to go.

The first snag hit us amidships. These tourists just weren't that easy to find. Forget the mystery snow leopard; suitable candidates to fill in our new survey were proving to be equally elusive. But through a combination of charm, threats and pleading, the tally started to mount, helped by the onset of the tourist season from mid-April onwards. Most people approached about it were actually more than happy to fill out the questionnaire. In the end, we bagged 406 of the blighters.

We didn't anticipate the second snag either: time. It was taking a lot longer to clean each filled-in questionnaire than we'd planned. Each consisted of two sheets/four sides of laminated A3 paper, and the answer to each question, written in permanent marker, had to be scrubbed vigorously with a wire sponge. One whole survey could be cleaned in 12 minutes. Five could be done in an hour. The 20 surveys we had with us took around four hours to clean in total. While my Nepali research assistants, Niki Shrestha and Rinzin Lama, looked after the household questionnaires, Maurice focused on this part of the research, getting through around 286 by himself before reinforcements in the form of myself and his girlfriend, Jorien Schumanns, arrived. After an initial three weeks getting the data collection phase of the project started in February in the Everest region, I had now returned to oversee the final three weeks of fieldwork in the Annapurna area.

The third snag proved to be the most significant. Our supplies of the cleaning solvent that we were using – methylated spirits – were disappearing a lot faster than we'd thought they would. Without them we reckoned we wouldn't be able to get the sheets cleaned properly. The initial batch of meths from Kathmandu simply evaporated, given the considerable altitude of 3,000 to 5,000 metres that we were working at. Resupplies from Rinzin did the same. Due to a miscommunication, Jorien and I brought 200 ml with us instead of the 500 ml requested by Maurice. Disaster. What were we to do?

They say necessity is the mother of all invention. So over the next couple of weeks, given the shortages of methylated spirits and with much begging, borrowing and stealing, we comprehensively tested the following solutions as to their efficacy for removing permanent marker from our tourist questionnaire: 1) Rectified spirits; 2) Saliva; 3) Hand sanitiser; 4) Coca-Cola; 5) Glycerine; 6) Rakshi (a local rice liquor); 7) Medical alcohol; 8) Bleach; 9) Salt water; 10) Soap; 11) Bagpiper Deluxe Whiskey; 12) Fresh water.

And after all that effort, and all that money, we concluded, having run out of all other alternatives, that fresh water and elbow grease could do the trick just fine.

As we sat for hours in various hotel common rooms – scrubbing, scrubbing, always scrubbing – the views from the office partially compensated for this part of our life that we would never get back. In Upper Pisang, on that still April day, we looked across the Marshyangdi valley and watched the war between these phenomenal forces that create and destroy mountains in slow motion. We heard the thunder. We felt the rumble.

And we also felt the wonder. These awe-inspiring forces in this awe-inspiring place did just that. They filled us with inspiration. Never have I witnessed such incredible power. Never have I felt so small. Never has the Sanskrit phrase from the Hindu scriptures been so imbued with meaning as then: 'In a hundred ages of the gods, I could not tell thee of all the glories of Himal.' I was so struck by its manifestation in this moment that I prefaced my PhD with it.

Yet amidst the splendour, and the many potential benefits of our research, the moral of the story was that innovation can have hidden costs. Maybe there was a reason this approach hadn't been tried before. I look back now, just over a decade later, and wonder why we didn't have it all set up as a survey on a tablet. Then again, done that way, we wouldn't have spent so much time witnessing the world-shaking power of nature in all its terrifying glory. Granted, the production of knowledge through research always takes time, but we hadn't properly anticipated just how much time this extra work would involve. As the old military adage goes, 'Amateurs worry about strategy, professionals worry about logistics.' Clearly we were still at the amateur stage – we hadn't given enough thought to the practical outworking of those bright ideas all those months back. Initially smitten with our quick-fix solution, we'd been reminded that whether it's snow leopard-friendly certified or not, whether it concerns ecology or economy, there's really no such thing as a free lunch.

Learned the hard way, it was a lesson of Himalayan proportions.

No free lunch

I learned other lessons here too, lessons that have helped me think through the complex economic case for and against the reintroduction of lynx, wolves and bears to Britain and Ireland. We'll unpack and apply these lessons throughout the remainder of this section.

Lesson one was the challenge of trying to value the intangible. As we've hinted at and discussed elsewhere in our story, it can be very complicated to try to value things economically that aren't normally valued that way. Or 'non-use values', to use the technical term.[2] Things like the wonder we felt as we watched those avalanches roll down the slopes of Annapurna III. Or knowing that there were snow leopards in the landscape around us, even if we never saw them. We call this existence value. Or believing that, in some small way, our research might make a contribution to ensuring this species lived on in this place far, far into the future. We call this bequest value.

Lesson two was that, despite the challenges and limitations, economics likes to try to value the intangible anyway. Our methodology – called 'Willingness To Pay' (WTP) – was based on trying to put a price on all these non-use valuations of snow leopards by the tourists visiting the Annapurna region.[3] It did this through something called contingent valuation. Contingent valuation creates a hypothetical marketplace for something and then ascertains people's willingness to pay for or forgo a product or service. In the case of our study, we gave people a short summary about snow leopards and the threats they faced, as well as the solution, in the form of a 'Snow Leopard Conservation Action Plan'. Survey participants were then asked if they would be willing to pay a snow leopard conservation fee – which varied randomly between US$5 and US$120 – to help implement the plan. The average figure, from the 49% of those surveyed who were willing to pay, was US$59.

Lesson three was that these methods have limitations. The WTP methodology has been heavily criticised for, among other things, 'hypothetical bias'.[4] This means that those being questioned know that the proposition they're being presented with is theoretical, and therefore respond theoretically. In other words, they're willing to pay theoretical money rather than the real stuff. If we'd whipped out a donation bucket during each of our 406 surveys and asked the individual to pay the sum they'd indicated, our average WTP figure would have plummeted. As it so happened, we found that the higher the snow leopard conservation fee proposed, the less likely people were to pay it, even theoretically. And then there were a whole host of implementation challenges that we went on to discuss in our paper.[5] Never mind the philosophical quandaries about trying to put an economic value on something like a snow leopard.

Lesson four was the extent to which tourists' visits to Annapurna were influenced by snow leopards. We asked them this very question. More than three-quarters said that snow leopards did not at all shape their decision to visit.[6] The rest varied between 'to a limited extent' and 'to a large extent'. Tourism with a cryptic species like a snow leopard is a tricky business. The species is very hard to see – I'm still waiting for my chance – meaning that the sort of mass wildlife tourism common in parts of Africa can't really be replicated here. For that reason we assessed the interest of locals and tourists alike in a 'snow leopard trail' of sorts, where people went on a guided tour into and through the places snow leopard frequently visit. They'd maybe even get to check a camera trap. But still the chances of a snow leopard sighting are small. And even for those who come for snow leopards, it's but one factor among many, including scenery, culture, physical achievement, and so on.

The final lesson was the uneven sharing of costs and benefits. Most of the costs of coexisting with snow leopards was and is borne by livestock herders, especially those mostly or fully reliant on their herds. On occasion, such as when a cat breaks into a night-time corral and succumbs to a killing frenzy, the results for individual households can be catastrophic. Nor is there much by way of a welfare state to provide a safety net, though there are often strong community networks of support that can step up in cases like these. Sometimes the official state livestock compensation scheme works. Sometimes it doesn't. Yet much of the value of tourism, and the limited amount of snow leopard tourism that there is, is captured by some very inspiring local tourism entrepreneurs, who have usually moved out of herding. What's more, parts of Annapurna on the tourist trail benefit much more from inflows of tourist cash than the off-the-beaten-track parts. There is a mismatch within *and* between the communities coexisting with the species, contrasting those paying the price with those reaping the rewards. To one group the snow leopard is a liability. To the other it is an asset.

The uplands of South-Central Asia may seem like a very long way from the uplands of Wales, Scotland, England and Ireland. Yet many of these same issues that I encountered in our study in Nepal apply here also. And as lynx remain the most likely candidate species for reintroduction, and have received the most attention in terms of economic analyses, there are a lot of parallels between this cryptic cat species and the ghost of the mountains. In particular, because tourism is often the keystone in the economic case for the reintroduction of lynx and other large carnivores to our islands, it is worth paying considerable attention to it. In fact, the economics of reintroduction in Britain and Ireland hinge upon tourism.

Overall, just like the continual tension between the forces that build and break down mountains that we were witness to in Nepal's Upper Marshyangdi valley, there is a continual tension between the economic costs and benefits of reintroductions. Because some of these costs and benefits are based on pricing things that are hard to price – as with our WTP methodology in Nepal – they can still leave, as with both the ecological and political cases, a lot of room for debate. And who will pay the price and who will capture the value remain key political ecology questions throughout. For whom will lynx – or wolves or even bears – be a liability? For whom will they be an asset?

As with large carnivore reintroductions, as with snow leopard conservation, as with just about everything else: there is no free lunch. There is always a price. And it will always be paid.

But by whom?

An independent report prepared for the Lynx UK Trust in 2015 comprises the most comprehensive attempt to model the economics of lynx reintroductions.[7] In terms of costs, the economic consultancy aimed to assess those relating to livestock losses, as well as the reintroduction project running costs, and any risks to human health from disease. In terms of the benefits, the focus was on recreation and tourism, deer control, and the existence value of lynx. They did this for two sites in England: Thetford Forest and Kielder Forest.

The costs first. The costs to human health, in terms of disease, were estimated at £0 in effect. In fact, the report speculated that lynx could actually help human health by reducing the incidence of deer-borne Lyme disease, a bacterial infection with ME/CFS-like symptoms.[8] The project operating costs over the five-year trial period were just under £1.5 million. This included things like capture and infrastructure costs (remember the wolf-holding pens we visited in Yellowstone in Chapter 11?), staff and research costs, and £185,000 for local education and consultation.

The estimated costs I'm most interested in are those concerning sheep losses. Based on a range of other studies from 20 European countries, the report estimated that each reintroduced lynx would kill 0.4 sheep per year, although with the Norwegian data of 10.5 sheep per lynx per year included, the average increased to 0.90. At a compensation rate of £140, twice the going rate for a sheep at the time, this resulted in annual compensation costs of £0 in the best-case scenario (assuming no sheep kills by lynx); £757 in the central scenario (0.4 sheep per lynx); and £5,378 for the worst-case scenario (2.84 sheep per lynx – the same as France).

I'm a great believer in hoping for the best but planning for the worst. Were the Norwegian scenario of 10.5 sheep per lynx per year realised, this would represent a figure of £19,883 per annum at Thetford and Kielder. Were we to apply this range of figures to the combined population estimate of 656 lynx for the Scottish Highlands and the rest of Britain that we considered in Chapter 11, we would arrive at £36,736 in annual compensation payments with the central scenario; £260,826 for the worst-case scenario; and £964,320 under the Norwegian scenario. We could assume a similar range for Ireland.

These are relatively small amounts of money. Even the nightmare Norwegian scenario costs pale when compared to the £150 million-plus and €74 million-plus spent each year on combatting bovine TB in the UK[9] and Republic of Ireland[10] respectively. Yet as we've seen time and time again on our reintroduction journey across Western Europe and North America, and as I've encountered in my British and Irish interviews, it's not just the literal financial loss that counts, though that is particularly significant for small-scale shepherds. It's the loss of valued bloodlines and prized animals. It's the perception of losses foisted on rural communities by urban elites. It's the symbolism of the loss of a way of life, dying by a thousand cuts, of which reintroduced predators are but one.

Furthermore, the Lynx UK Trust assessment doesn't make any mention of deterrence costs. If prevention is better than cure, and a pan-European review of livestock protection and compensation schemes suggests that it is,[11] then these should also be factored in. Think employing staff, perhaps through the various agricultural colleges across reintroduction regions, to train farmers to use the range of livestock protection tools we saw in Chapter 6: fencing, guard animals, collars, corrals and increased human presence. Think also paying for all these tools, the human option especially. And think, if farmers were to be paid for the time involved in adapting their farming practices to minimise losses, how much that might add up to. This is the sort of detailed economic analysis that this debate needs but doesn't seem to have yet. Here and for now, the best I can do is pluck out the £2.76 million spent by the Swiss government in 2020/21 on livestock protection methods and advice, and throw it into the mix.[12] With the caveat that the majority of this is spent on wolves rather than lynx. Factoring in wolf and bear losses – of sheep but also, to a lesser extent, cattle and horses – would add whole new columns in our cost–benefit analysis.

The political ecology perspective that we employed in the last chapter noted the influence of economics and power dynamics on environmental change, as well as the interconnection of all three elements.

If we look at these cost estimates through this lens, then it becomes apparent that we can't take the reasonably minor financial sums in isolation. Rather, we also need to consider them in relation to the panoply of real and perceived non-financial costs incurred, especially by those living closest to lynx, or other reintroduced carnivore species. The trouble is, as we saw with my snow leopard research in Nepal, some of these things are very hard or impossible to value economically. Meaning that a lot of the costs of lynx reintroductions are likely to be subjective and, by default, political. And philosophical – which is what we'll explore in Chapter 14. It's all rather complicated.

There is definitely no free lunch available here.

The cat that got the cream

Next we consider the benefits. Again we use the 2015 economic assessment prepared for the Lynx UK Trust. The value of the reduction in deer was estimated at £3,359,786 across both Thetford and Kielder forests over 25 years. This assumed lynx preying on muntjac and roe deer primarily, with the occasional young red or fallow. It factored in a reduction in traffic accidents from collisions with deer, as well as reduced damage to forestry operations.

The study also included a WTP methodology to estimate the existence value of reintroducing lynx. In other words, the value of the species being in Kielder or Thetford forests even if there were no deer control or tourism benefits. Although it didn't gather data directly from respondents around the two sites, the report extrapolated using a figure of 60% support for a trial lynx reintroduction within the next 12 months from the Trust's representative survey, from another study that we considered in the previous chapter.[13] This isn't a million miles away from the 49% of respondents in our Annapurna survey who were willing to pay towards snow leopard conservation. The WTP was set at £2.76 per household per year, based on a review of similar studies from Europe and North America. This gave a total figure of around £14 million over 25 years. The worst-case scenario modelled was £1.2 million, while the best case was £282.8 million. Tellingly, given the limitations and critiques of this methodology that we discussed earlier, this existence value estimate was not included in the final cost–benefit analysis table in the report's conclusion.

What was included was a figure of £65.7 million for the recreation and tourism benefits at both Kielder and Thetford. It extrapolated from a review of other wildlife tourism studies of birds and mammals across Britain. It factored in the five-year trial period, as well as the longer term. The methodology also tried to take into account the extent to which people visited because of lynx, and the extent to which their expenditure in the local area could be attributed to lynx tourism. Finally, the analysis estimated that 27 full-time equivalent jobs would be supported.

These are considerable sums. Given, as we discussed in Chapter 11, how often the ecological case for reintroducing large carnivores to Britain and Ireland is benchmarked against the Yellowstone wolves example, it's worth drawing on some of the economic figures from here too. We can also use these to assess whether the projections matched actual benefits. The 1995 Yellowstone wolf reintroduction was projected to generate US$27.7 million in tourism revenue per year.[14] Ten years later it was estimated at US$35.5 annually.[15] After another decade or so, it had risen to US$45.5 million in yearly wolf-related tourism spend. Here, though, livestock compensation costs were more than double what was projected, though this was due, in part, to the fact that wolf population increased and expanded quicker than anticipated.[16]

Overall, when the Trust's economic analysis weighed up all the costs and benefits that we've discussed, it projected that for every £1 of cost there would be £47 of benefit, though it did acknowledge that this was a preliminary and exploratory study. If this ratio is accurate, then the lynx looks very much like the cat that got the cream. From it, though, four things stand out that inform our broader enquiry into the economic case for and against large carnivore reintroductions.

First, the case for tourism especially, and to a lesser extent deer control, appears to be strong. Although it is hard to predict the future, the levels of public support for reintroductions that we considered in the previous chapter are likely to translate into willingness to actually pay for lynx tourism experiences, even if the animal itself is rarely seen.

Second, coexistence costs are underestimated. Again, rates of livestock loss are hard to predict. At best it could be the middle scenario of

0.4 sheep per lynx per year. At worst it could the Norwegian rate of 10.5 sheep per lynx per year. Wolves especially would likely add significant additional compensation costs, not least because of their capacity to take larger farm animals. And the most significant coexistence costs of all – paying for the time of farmers and their advisors to adapt their agricultural practices to include a suite of deterrence methods – would add considerable cost again.

Third, costs and benefits are unevenly spread. This is something that has also been noted with a multi-agency study of a beaver reintroduction trial in Devon.[17] Those paying the price of coexisting with reintroduced lynx and other predators, specifically livestock farmers, are unlikely to be the ones gaining the benefits of the tourism bonanza that may occur. Evening up this equation is absolutely key to making projects like this work, such as through the coexistence-focused funds that we considered in Chapter 7. Unless lynx, wolves and bears become assets to a significant majority of landowners and landworkers, instead of liabilities, the allocation of costs and benefits will be skewed. Social resentment with political implications could follow.

Fourth, much remains subjective. Willingness To Pay methodologies remain an imperfect tool to value many aspects of predator presence. And as with the ecological benefits, the full economic benefits, and costs, are unlikely to be fully understood for years or even decades. What's more, many of the costs and benefits cannot be computed or quantified or enumerated. For that, we look to philosophy to assist us. But in the many subjective grey areas, and among the range of worst-to-best-case projections, there is much territory for political animals to fight over.

But irrespective of the precise ratio of costs and benefits, or their precise allocation across urban and rural areas, or across the farming, forestry and tourism sectors, large carnivore reintroductions are going to cost a lot of money. How can they be funded? Just as we've gone through the micro-economics of reintroductions with a fine-tooth comb, it's worth zooming out to consider some of the macro-economic trends that will shape the likelihood of this idea being successfully implemented in the coming decades.

Start with the costs of funding reintroduction pilots. If we take the Lynx UK Trust figure of £1.5 million as the typical cost of a carnivore reintroduction pilot in Britain or Ireland over five years, then this should be easily within the financial reach of both medium-to-large charities and wealthy landowners. For the charities, it's also an extraordinary fundraising opportunity. Having been a charity fundraiser, and having also taught fundraising at postgraduate level, I've become convinced that fundraising is all about storytelling. And what a story the return of the lynx or wolf or bear would be, especially if you live in an urban jungle, only visit the reintroduction landscapes on holidays, and don't own any land or livestock. As exciting a fundraising opportunity as this may be, however, it could risk exacerbating the rural/urban or local/outsider splits that keep appearing as an issue in this debate.

Similarly, there is the option of accessing public instead of philanthropic funding for projects. In particular, many European large carnivore projects over the last few decades have been funded through the EU's LIFE programme.[18] Ongoing access to this deep financial pool gives the Republic of Ireland an advantage in this regard and puts the United Kingdom at a disadvantage.

However, the real challenge lies in funding the costs of coexistence beyond the lifetime of a five-year pilot. As a population of, say, lynx spread over the Scottish Highlands, the west of Ireland, South East England or mid Wales, the bill for compensating for killed or injured sheep will rise over time. But it will probably be dwarfed by the cost of equipping and training farmers in the deterrence methods required to protect livestock from predation. As we've seen, these costs will rise significantly if bears and especially wolves are involved. If we take the somewhat arbitrary Swiss figure of £2.76 million for funding coexistence costs annually, and apply this in an Irish or British context, who is going to foot the bill? And not just for the first five years, but for the next 25, 50 or even 100 years?

Even taking into account the relatively strong economic case for reintroductions, especially with tourism, without the resources of the state, charitable and private sector funding sources will struggle to adequately resource an endeavour of this scale in the long term. And while there is a clear trend towards public investment in public goods like biodiversity, as discussed previously, securing enough long-term public funding to offset the costs of coexistence with reintroduced large carnivores is likely to be a challenge.

It's not that the British or Irish states can't afford it. In 2023 the UK government's budget was £582.6 billion,[19] while the Irish government's budget was €90.4 billion.[20] It's more a matter of priorities. Spending on nature conservation just doesn't have the same political appeal as spending on health, education or defence. In fact, the UK faces a funding gap of £44 billion to £97 billion to meet nature-related outcomes over the coming decade.[21] Meanwhile, in the Republic of Ireland, the entire budget for the National Parks and Wildlife Service was €11 million in 2017.[22] In the same year, the greyhound- and horse-racing industries received €16 million and €64 million in public funding respectively. Trade-offs and competing priorities for public funding – think horse trading rather than horse racing – will always exist. Do large carnivore reintroductions have the public and policy support to muscle in on this long list of expenditure priorities? Time will tell.

Time will also tell how changes to agricultural subsidies and payment schemes across both Britain and Ireland change upland areas, the very places where reintroduction projects are most likely. Facing the stiff prevailing winds of ageing farmers, limited economic opportunities, rural depopulation and a changing climate, upland areas may shift towards larger-scale extensive sheep farming, default rewilding through land abandonment, or what is called High Nature Value farming.[23] The last of these is sometimes called farming for conservation, or even sometimes rewilding, of a sort. These last two options are likely to increase the likelihood of large reintroduction projects, while the first one would decrease the feasibility. Yet in these landscapes, and in the lowlands as well, if compensation payments for carnivore kills are taken from the same limited funds as other forms of agricultural subsidies or payments, farmers and landowners will not be happy. And there will be trouble over it, as multiple interviewees in the Netherlands explained from their country's experience.

In mathematics and machine learning there actually is a 'No Free Lunch' theorem.[24] It states that 'no one optimum optimization algorithm exists'. In other words, there may be no one-size-fits-all solution for funding large carnivore reintroductions. Each approach may have merits of its own and reach the same result in the end. The potential tourism benefits of a given reintroduction project may depend on its proximity to urban markets, for instance. At another site, the potential livestock compensation costs may depend on the number and scale of landowners – a few big ones with cattle or lots of small ones with sheep. Funding schemes, whether public, private or philanthropic, or a mixture of all three, will need to be tailored to the unique economic requirements of each place. And innovative financial schemes, like the 'Payments to Encourage Coexistence' fund proposed by Amy Dickman and colleagues,[25] or Colorado's Wolf Conflict Reduction Fund,[26] both of which we considered in Chapter 7, will be crucial to ensure that economic benefits accrue to those paying the economic price.

As with our valiant efforts to trial different questionnaire-cleaning solvents in the shadow of Annapurna, necessity is likely to be the mother of all invention here too. With a will there's a way, and with adaptive management in place, many of the challenges that reintroduction projects inevitably face can be solved with a 'the answer's yes now what's the question' attitude. But it's one thing taking the coexistence risks that predator reintroductions entail with your own land and livestock if you own 100,000 acres in the Scottish Highlands. It's quite another thing entirely if you're taking the same coexistence risks on a patchwork of land that involves someone else's small- to medium-scale family farms – who are just about keeping the show on the road.

Similarly, the trial-and-error nature of reintroduction projects, even despite the best planning and projections, requires a certain degree of political and social latitude to operate within. In the swamp of politics where emus roam, and with the harsh and unforgiving scrutiny of social and traditional media, projects that take risks and fail may find that this social and political operating space disappears very quickly. Even at the planning stage, this can have knock-on effects for similar projects and affect the likelihood of their success. In the case of the reintroduction efforts of the Lynx UK Trust, whose admirable cost–benefit analysis belied an overall lack of support and consultation,[27] the debacle may have set back lynx reintroductions by at least a decade, as we previously noted and as several people pointed out to me. The cat did not get the cream in this particular case.

Before I set up Jubilee Farm, my then Chief Financial Officer-uncle told me that any project takes twice as long and costs twice as much as planned. In my optimism bias, I ignored him and carried on. But, in the end, he was right. In the end, large carnivore reintroductions are likely to cost more than the cost–benefit analyses produced by and for NGOs with rewilding optimism bias. In the end, beyond limited pilots, they are also likely to require the financial resources of the state to maintain an acceptable cost–benefit ratio over decades. This makes them a profoundly political issue.

It also makes them profoundly moral and ethical issues, as we shall see in the next chapter. What price should be paid for large carnivore reintroductions, or for that matter, snow leopard conservation, is not a question economics alone can answer.

But economics does matter in this debate. A lot. As the intimate and ancient balance between the natural economy – ecology – and the human economy unfolds around us all, the defining issue again and again is of synchronising the *oikos* – or 'home' – of ecology with the *oikos* of economy. Only through striving for this can we balance the equation between cost and benefit, use and non-use values, for both large carnivore reintroductions and for snow leopard conservation. Not to mention climate, biodiversity, food security and human flourishing. And whether our home is beneath the icefalls of the Annapurnas, amidst the lochs of the Scottish Highlands or in the housing estates of Limerick or London, only through striving for this balance can each of these places be, always and forever, home sweet home.

No matter that it is a task of Himalayan proportions; as elusive as a lynx or a snow leopard it may be; and even if there is no free lunch to be had anywhere: this symphony of synchronisation must be the soundtrack of this century for our one and only home.

It is the sound of our planet living.

CHAPTER 14

For the Love of Wisdom

Atop the spire sat a pumpkin. The magnificent university hall dominated the oval campus green in front of it. Its red-brick construction reminded me of the similarly imposing main building at one of my alma maters, Queen's University Belfast. The copper-green roof atop the central tower, a feature of several Belfast landmarks, further reinforced the Ulster connection for me. But that's where the similarities ended.

'It's a tradition here at Halloween to stick a pumpkin on top of the main university building,' noted Christopher, as he gestured upwards.

I looked up at the central tower as it soared into the sky, and swallowed nervously. Despite my love of mountaineering, I had a slight fear of heights, which seemed to be getting worse as I got older. The exposure for a climber trying to ascend this building was considerable, never mind for one carrying a pumpkin.

'Is there an access door at the base of the copper roof or do they climb the tower from the main part of the building?' I wondered.

'No one knows,' laughed Christopher, 'but every year at this time it appears on top of the spire.'

The pumpkin pointed to the changing of the seasons. Autumn was setting in across the Mountain West and I felt it here. The hills were alive with its colours and at night, the temperature was dropping. The stars sparkled above my campground in the woods outside town, on the banks of the Blackfoot river. At the north-western apogee of my USA road trip, I had come to Missoula, Montana.

Missoula was home to the University of Montana (UM), and its Halloween pumpkin. It was also home to Christopher Preston. Christopher was an environmental philosopher and writer, and I had come to philosophise with him about the ins and outs of large carnivore reintroductions to Britain and Ireland. British himself, he had come to the USA temporarily and stayed permanently.

As we basked in the still-warm October sunshine, we looked out on the green space at the heart of the university. The tops of the young trees ringing its perimeter were flushed red. Students lounged in wooden deckchairs. Others read. Still others casually threw a football. So far so like the green spaces at the heart of Queen's in Belfast, pumpkins aside. But there was one major difference, as Christopher was keen to point out to me.

'This is where the bear was yesterday evening,' he said, as he pointed at the quintessential college scene.

Christopher had shared the emergency alert email. The university was legally obliged to issue emergency notices like this when *an ongoing dangerous situation exists on or near the UM Campus*. As well as the location of the female black bear and her cubs, the short missive gave the usual sensible advice that I'd encountered from central Colorado to north-west Montana: remain calm; don't approach; don't run; be loud and intimidating if the bear comes towards you.

But there had been no panic among the student body. No faculty letters calling for the cancelling of bears. No complaints from the residents of this small city. In fact, no one batted an eyelid. Life simply carried on, albeit 'bear aware'. That's because the bear on campus was nothing new. Bears were an accepted part of life on the outskirts of, and sometimes even in, Missoula.

'Last fall the berry crop failed in the surrounding hills and black bears came into the city to look for food,' Christopher explained. 'At one point, there were 150 bears in Missoula. No one was injured.'

Christopher was a keen mountain biker. Although we'd met at his house for tea first – and amidst my complaints about the quality of tea in North America, he'd produced some of the good stuff from home – we had taken to two wheels to see the sights of this charming college town. Like Fort Collins at higher latitude. And with a lot more bears.

'You must have seen some up close while biking?' I asked, as we pedalled slowly through tree-lined streets.

'I've seen about forty bears over the years, but all of them were "leaving" when I encountered them. And I've never even seen a mountain lion.'

I remarked that I'd never seen a leopard either in all my years walking and running through the mountains of Malawi. Or a snow leopard in more recent times in Nepal.

Christopher continued: 'When I moved to the USA, I was that person who thought I'd be attacked and eaten. For ten years, I was afraid to go out into the street.'

It's not that bears can't pose risks to people, especially brown ones. The 'bear aware' guidance and bear-proof bins I'd encountered on my travels were attempts to minimise these. As were the bear spray I'd carried and the bear protocols I'd observed when in Yellowstone and the Tom Miner Basin, where the news was of a spate of recent grizzly attacks in and near Montana. And I still remembered how little time I'd had to respond to Robobear when it charged me in Cody, Wyoming. All of these had only cemented my utmost respect for bears. But still the statistics spoke for themselves: more dangerous creatures than even grizzly bears roam our streets.

I shared how Paula and I had been charged by an XL Bully and how this had been, far and away, the most frightening and dangerous animal encounter I'd ever had. And as we've seen in this book, I've had more than my share of interesting animal experiences. Christopher nodded and explained how his wife had been attacked by a pit bull while gardening in the summer of 2022. She had been seriously injured, requiring hospitalisation and 25 stitches.

Despite that awful experience, he made the same point that he'd made in a recent article about what is and isn't wild:[1] 'About thirty to forty people are killed by dogs every year in the US – but we don't want dogs banned. The levels of fear in the UK about large carnivores are a gross distortion of reality.'

'Why is that?' I wondered in agreement.

'Because, in part, Europe is a culture obsessed with civilisation, and it fitted well to banish these animals from our human geographies. In other words, the wild is out there, not over here.'

It is a philosophical idea at the heart of the reintroduction debate. It is a philosophical idea that shapes the history of the wild as a concept and as a place, as we observed in Chapter 3. It is also a philosophical idea that shapes the history and process of rewilding, adding, as we discussed in Chapter 4, a temporal and procedural element to the conceptual and spatial 'wild'. And as we've seen throughout our journey together, it is an idea that elicits strong, frequently contested and always complex feelings in us as humans.

But love it or hate it, the wild defines reintroductions. Embrace it or fear it, the wild dominates discussions about the return of lynx, wolves and even bears to our islands. What is wild and where is wild would also underpin our conversations as we cycled around Missoula.

For Missoula is a wild place. Even more than Fort Collins, with its urban forester and its occasional itinerant mountain lions wandering into town from the foothills of the Rocky Mountains, Missoula was surrounded by wilderness. North of the town, almost to the edge of the university, a series of national forests stretched all the way to Glacier National Park on the Canadian border. To the south, the Selway-Bitterroot Wilderness was the largest wilderness area in the lower 48. Earlier that day, I'd gone with Mitch Doherty of Vital Ground, a conservation land trust focused on protecting grizzly habitat, to see a particularly strategic conservation corridor that they'd recently purchased. A single underpass under Interstate 90 secured access for the grizzly population expanding southwards towards Missoula and into this vast new area beyond it.

And in this wild place, cycling was the perfect way to consider the philosophy of wild in all its nuance and complexity. Nietzsche argued that 'all truly great thoughts are conceived while walking'. It was true: I had worn through at least one pair of shoes thinking through this book. Yet I hoped that truly great thoughts could also be conceived while cycling. The bilateral stimulation of both these motions help the twin hemispheres of our brain,[2] and arguably the 'parts' of our multiplicitous minds, to process information. Philosophy, being one of the five key factors influencing large carnivore reintroductions to our islands, is where the boundaries of the lived realities of our outer landscapes and the individual and social psychologies of our inner landscapes overlap most fully.

As we cycled, I peppered Christopher with the various philosophical propositions I'd thought about or heard from my interviews and visits. As we weaved between cars and joggers and fellow cyclists, he responded. Trying to stay on the correct side of the road, and in the saddle, I supplied the counter-arguments. The wisdom of rewilding unfolded on our dialectical perambulation. Our inner and outer landscapes fused.

The broadest of minds

From the college oval, we cycled across a car park. A steep trail led straight up the side of the adjacent hill. Fit-looking students were limbering up for a challenging hike or run. Beyond that, a gravel trail led along the side of a sports pitch. On one side was a chain-link fence. On the other side, forest.

'This is where bears are most commonly sighted,' shouted Christopher, as he gestured rightward to the forest. I tried to stick close enough to his rear tyre to hear him but not close enough to collide if he stopped suddenly.

He continued: 'There's an argument that large carnivores belong on the land, even after absences of long periods of time.'

Gesturing leftward to the chain-link fence, I supplied a hypothesis that I'd heard frequently in my interviews with British and Irish farming unions, as well as with Swiss and Dutch farmers: 'But if the land has changed considerably since these species were last present, especially through farming and the presence of a much higher human population, surely conditions are no longer right for their return?'

'That's a valid point,' noted Christopher, 'but I wonder what gives farmers the authority to make the argument as being more valid than others?'

'I suppose they are the ones on the frontline of carnivore coexistence, and stand to lose the most from their reintroduction. Also, how far back in time should we go to restore this "belonging"?'

Christopher had recently been back in the UK to visit his elderly parents. 'When I was in England recently I went to visit the bison reintroduction site in Kent. There haven't been bison in England for over a thousand years. But they restore ecosystem integrity and function to those woodlands. I think that is the strongest case for the reintroduction of predators, as George Monbiot points out.'

I talked about my recent trip to Yellowstone's Lamar valley with Joanna Lambert to understand the ecology of wolf reintroductions, and how I felt that the 'wolves-change-rivers' story was a bit overblown.

'But Monbiot also makes the point in *Feral* that having wolves return makes landscapes more exciting,' I supplied.

By now we'd reached the end of the narrow gravel trail alongside the sports pitch. We came out onto a broader trail that ran next to a river. Christopher pointed at the open hillside on the opposite side of the river.

'Just last winter, someone saw a bobcat hunting over there. Landscapes are definitely more alive and exciting with animals like this present.'

I concurred. As beautiful as they are, walking and running through the British and Irish countryside never holds quite the same excitement as I experienced in Malawi, Nepal or North America.

As we paused on the track beside the river, a steady stream of walkers and runners passed us in both directions. Gesturing to them, Christopher explained his argument that 'people, and arguably a majority in Montana, want large carnivores on the land'.

'I get that,' I responded, 'but to what extent can the will of the majority be enforced on a minority who don't want it but have to live with it – like the wolf reintroductions in Colorado?'

'That's also a valid counter-argument,' conceded Christopher, 'and minorities' concerns do need to be taken into account. But the democratic process is the fulfilment of the majority's wishes.'

As we left the riverbank and pedalled back into town, our philosophising moved from the abstract to the concrete. Christopher thought that the issue of financial compensation was central to the debate, in Colorado and elsewhere. He even articulated a willingness to pay higher taxes to fund this, the first person in the USA I'd heard this from. It must be because he's British.

'But,' I countered, 'in making money a central part of the process, is there a danger that the relationship between people and predators becomes monetised, as in the Michael Sandel market society critique?' Michael Sandel was a philosopher who critiqued the expansion of market forces into various parts of society, including our relationships with nature.[3]

Christopher paused thoughtfully as I pulled up alongside him. We were almost back at the car park.

'Maybe,' he ventured, 'but, pragmatically, people's economic needs have to be met to facilitate coexistence.'

On the same note, I wondered whether putting sensors and collars on large carnivores to facilitate coexistence, as I'd seen in Wyoming and as technology was likely to enable more of in the coming decades, was 'dewilding' the wild?

We were now back in the car park at the foot of the steep hiking trail. We hopped off our bikes to cross a busy road. Beyond it was the main University of Montana football stadium.

Christopher said that his environmental philosophy class had looked at this very issue over a number of sessions. He continued: 'Again, pragmatism needs to win the day: much if not most wildlife is changed in some way through interactions with people. Like those wolves you saw in the Lamar valley, who are more comfortable with people due to the sheer numbers of tourists viewing them.'

We paused for a selfie at the main entrance to the stadium. The college team was the Montana Grizzlies. A huge billboard showed the dates of their fixtures in the coming semester, as well as their opponents' mascots. Of the six, four were predators: two bulldogs, one tiger and another bear.

It reminded me that Belfast's nineteenth-century coat of arms had a wolf on it. Then I mentioned that I'd done a live show at the Northern Ireland Science Festival on the history of our fascination with big cats, and how they and other large carnivores had come to symbolise power in various ways, including in the sporting world.

'The number of NFL teams that have a big cat as their mascot is incredible,' I enthused. 'And in the UK you have the Premier League in football, the British and Irish lions in rugby. Even my native Ulster play in the Kingspan stadium – a leaping lion is the symbol of their main sponsor!'

The conversation moved away from the unfamiliar world of American football to the more familiar terrain of rugby. With the World Cup on, I'd been starved of rugby conversation for the last three weeks. Christopher congratulated me on Ireland's performance to date and we scrutinised England's. A big showdown with Scotland lay ahead for Ireland and I planned to tune in as I drove from Missoula to San Francisco.

As we got back on our bikes and cycled round the stadium, we both agreed that predators had a profound place in our psyche – and in our philosophy – whether we loved them or feared them. Or maybe a bit of both.

This significant symbolism, of these and other species, was on display in a visceral way at our next stop along the river. The Boone and Crockett Club was a famous hunting society headquartered in Missoula. We poked our heads through the door. Champion trophies lined the walls of the foyer, from deer and wild sheep species to musk ox and pronghorn. Bear and mountain lion were here too, frozen in repose. I admired their beauty, even in death. Here too these animals represented power: in their own right, and the power of humanity's ability to take life.

We headed back to the house, passing a restaurant dumpster where bears had frequented the previous autumn. We turned away from the riverside trail and back onto the leafy streets, lined with wooden houses of colour and character. As we neared the end of our circuit of the city of Missoula, we also came back to the start of our metaphysical journey: what and where is wild?

'Since I moved here, I've come to appreciate the long, long history of Native American coexistence with nature, including large carnivores. It challenged my preconceptions about nature as truly wild or pristine. Indigenous peoples were interacting with and managing nature throughout the pre-Columbian period of American history.'

I nodded. In Malawi, in Nepal and even just a few days previously in Yellowstone, I'd seen protected areas like national parks that, on one hand, were held up as models of this idea of 'pristine' wilderness and yet, on the other hand, either still had – as in Annapurna – or once had – as in Yellowstone and most of Malawi's parks – people living in them. The exclusion of indigenous people from these places was often part of a chequered colonial past.[4] Yet in our chequered post-colonial present, with the full range of human activities placing a stranglehold on the survival of many species, places like these are essential for the future of wildlife. They provide core spaces where species like lynx and wolves and bears are relatively free from human interference, aside from the muted click of camera lenses and the buzz of tourist vehicles.

For this very reason, though, I had spent very little time in national parks and other protected areas on my journey. For me, it was the messy and contested spaces outside these core wildlife zones where the real conflict over carnivores lay. It was these messy and contested spaces where our coexistence toolkit – deterrence, finance, force, enterprise and governance – was needed to manage the myriad relationships between people and predators, and between people over predators. And it was in these messy and contested spaces where wild was objective fact – because wildlife was everywhere we were or had been, from 'pristine' parks to sprawling metropolises. But this was also where subjective perspective came into play – which point on this continuum was wild enough for large carnivores?

For some, the mountain fastness to the north and south of Missoula is where these animals belong. For others, the edge of, and even in town is just fine, provided sensible precautions are taken. Still others would prefer if predators, especially wolves, were contained to national parks alone. And if there is a strong degree of subjectivity about where wild is, there is also a similar spectrum of belief about what wild is. If wild is, as Christopher discusses in his article on dogs and bears,[5] 'threat', as it has long been perceived, especially in European thought, then surely the pit bulls and the XL Bullies of this world are wilder than the black and brown bears of Missoula? And in human terms, is it not the irresponsible and usually urban owners of dangerous animals like these, who let these specimens roam the streets without leash or muzzle, who are wilder than the Monbiots and Prestons of this world, half-feral in the landscapes they love?

Wild is a slippery concept philosophically. In our cycle tour of Missoula, Christopher and I had only skirted around the edges of this and multiple other tenets that underlay the reintroduction debate in Britain and Ireland. For every argument, there was a counter-argument. For every perspective that thought parts of Wales, Scotland, England and Ireland were wild enough for lynx, and maybe even wolves, there was a perspective that thought they were not. For every opinion that felt our ecosystems would be more complete and more dynamic with the return of these animals, there was an alternative opinion that felt they had been extirpated for a reason. And should remain so.

I was back at the beginning of my journey. In literal terms, we had arrived back at Christopher's lovely house, the trees on the broad street casting shadows on my car and on the garden as the afternoon lengthened. In symbolic terms, I was back to thinking about the idea of spectrums that I first articulated in Chapter 2: spectrums of opinion within our psychic multiplicity, shaped by our individual psychology; spectrums of opinion between us, shaped by our social psychology. In my own internal debate about the case for and against large carnivore reintroductions, I could still appreciate and feel the tug of contrasting philosophical perspectives. In the external debate that plays out in the decades to come, these same positions will be discussed and dissected at length, as they should be.

To me, it seems that there isn't really a single philosophical case either way, but multiple philosophical cases. From multiple angles. Perhaps the philosophical case that matters most is being open to each other's points of view. And accepting that these issues are varied and complex, difficult to boil down to simplistic and comforting categories of for/against or us/them. Perhaps wisdom is the best philosophical approach to take: 'the ability to use your knowledge and experience to make good decisions and judgments'.[6] Philosophy, after all, literally means *the love of wisdom*.

My mind, though, had been broadened by my bike tour of Missoula with Christopher. Truly great thoughts can be conceived while cycling, it turns out. I was wiser for it, even if I was still unsure of what final philosophical conclusions I had come to. Maybe I never would come down on one side or the other. Maybe it was OK to hold a spectrum of beliefs on this and other issues.

And in the back-and-forth dialectical between us, I saw the benefit of this active philosophy approach. If I could take everyone I'd interviewed in Britain and Ireland, farmer and rewilder alike, on a bike tour of the Netherlands, we could hash out an approach to managing and governing coexistence for carnivore reintroductions as we went. Once we'd tackled the Dutch countryside, we'd be ready for some Swiss mountain biking. Then the USA's Mountain West. By the time we'd finished, not only would we all be very fit, but we'd have come to an understanding, I would hope.

That broader minds are always better. That wisdom is a treasure without price. And that, wherever we are on the (re)wild spectrum, philosophy is key to understanding ourselves, each other and our place in the world.

The largest of hearts

Four months earlier I'd been in Nijmegen, a small and historic city in the mid-eastern Netherlands. Despite being in such a busy and crowded country, Nijmegen was in many ways the rewilding capital of Europe. A number of national and continental rewilding organisations were headquartered here. Remember the pioneering Dutch projects of Zuid-Kennemerland in Chapter 3 and Oostvaarderplassen in Chapter 5? Much of the innovation and entrepreneurial dynamism that underpinned them emanated from places like this.

I had several meetings scheduled in Nijmegen, including with Martin Drenthen. Martin was an environmental philosopher at Radboud University who specialised in the philosophy of rewilding. A friend of Christopher's, he'd actually introduced me to my Missoula host. Before I interviewed Martin, he'd invited me to a lunchtime seminar introducing an exciting new project to his colleagues in the philosophy of science department. WildlifeNL was an eight-year, multi-million-euro project led by three universities, including Radboud, in collaboration with a range of NGO partners, including conservation, hunting and farming organisations.[7] Its stated goal was 'dynamic co-management of human–wildlife interactions in the Netherlands'. In other words, finding pragmatic solutions for coexistence between people and wildlife. Zuid-Kennemerland was one of WildlifeNL's two 'living labs', in which solutions that changed the behaviour of both humans and animals to enable coexistence were tested and refined.

What I found so fascinating and inspiring about the project, as Martin introduced it in the crowded seminar room, was its interdisciplinary nature. Not only were there parts of the project focused on data and technology, but also on ethics and sociology. An app was being designed, but with input from philosophers like Martin. I was delighted to see an entire work package devoted to 'adaptive governance'. It was a pioneering project in a pioneering nation. It was coexistence research at its best, bridging gaps between the sciences and the humanities, between theory and practice and between different interest groups. And like my cycling conversation with Christopher, it disrupted and challenged the notion of what and where wild is. Not just in Missoula, where the forest stretched northwards, uninterrupted, to the Arctic Circle. But also in the Netherlands, where a population three times that of Ireland's squeezed into half its area.

My discussion with Martin continued to broaden my mind. His specialism was environmental hermeneutics. I nodded politely, knowing I'd heard the 'hermeneutics' bit before from a theologian friend, and trying hard to remember what it meant from my previous incarnation as a student of the medieval era. Hermeneutics, it turns out, focuses on interpreting and understanding texts. Environmental hermeneutics expands this to include interpreting and understanding landscapes.[8] Like a book of wisdom, or even like those sporting mascots on the stadium billboard, these are all imbued with implication and symbolism that give meaning to our lives. Martin expanded on two particular concepts that can help us to understand each other's philosophies and, in turn, our various philosophical cases for and against returning apex predators to our islands.

The first of these ideas is 'moral experiences'.

Moral experiences are seminal and formative moments of shared moral clarity in a society, Martin explained. In relation to the natural world, this meant less 'the experience [of nature] of different groups of people' and more 'different aspects of the moral experience of nature in which all of us are able to discern certain meanings and take them seriously'.

He had done considerable work on understanding how these moral experiences varied between rewilding and traditional conservation, with its nature reserves, management plans and hands-on approach, like that tiny patch of land that we accidentally let a herd of cows into in Chapter 5.

'The moral experience of traditional conservation is "nature is vulnerable" and requires us to take care of it,' noted Martin 'For rewilding it's more that "nature is resilient" and requires a respectful attitude of self-restraint on our part as humans.'

'The concept of "self-willed nature" – that nature can take care of itself?' I ventured.

Martin nodded and continued: 'It's possible to find a balance between these two narratives – for instance, that nature can be both fragile and resilient – by focusing on areas of overlap and commonality. The care perspective can correct the dangers of rewilders being morally indifferent towards vulnerable species disappearing as a result of natural processes. A rewilding perspective can correct those nature managers that forget that caring for nature should not lead to "gardening".'

Now, much like me and especially through the WildlifeNL project, Martin was using the same framework to consider farmers' perspectives on rewilding and wildlife coexistence, including their moral experiences.

'This is what I call the "farmer's ethic", which recognises that some places are fertile enough for us to be able to grow food, and sees this as a "gift" of nature that we should appreciate. This experience of fertility calls for an attitude of gratefulness, and should motivate a dedicated working of the land.'

I suggested a 'nature is functional' moral expression for farmers, but also that many would recognise nature as being fragile and resilient as well.

Martin concluded: 'This third ethical demand creates a tension with the other two moral experiences. The challenge is to find a middle ground. In other words, the various forms of landscape ethics should be understood as complementary perspectives, all of which should, in their own way, inform our moral attitudes towards nature and landscapes.'

I could see these different moral experiences displayed in the conversations I'd had across Western Europe and North America. I could see them in the array of groups with an interest in large carnivore reintroductions to Britain and Ireland. And I could see all three of them in myself. Eighty thousand words into writing this book and the tension between them was still there.

Moral experiences are also shaped by the meta-narratives, or overarching stories, of their time. In terms of Britain and Ireland, we can divide our history into three very rough and broad periods: pre-Christian/pagan, Christian, and post-Christian/secular. How did, and how do, each of these shape the moral experiences of people who lived and live through them, especially in relation to large carnivores? As with the nuanced philosophical arguments Christopher and I batted back and forth on our bikes in Missoula, there is nuance here too.

It is tempting to ascribe to our pre-Christian era a vision of romanticised pagan societies dancing naked amidst sacred groves in communion and harmony with nature. There was certainly veneration of trees, animals and other natural features that were part of a more ecocentric, or nature-focused, worldview.[9] Yet much pre-Christian mythology in Europe displayed a deep-seated fear of untamed nature: in Ireland, for instance, the image of a hero conquering wild nature in the hunt was a common motif in our legends.[10] As we discussed in Chapter 2, nature, including big predators, equalled fear as well as wonder.

In the same way that we can romanticise pre-Christian perspectives on the natural world, there is also a temptation to simplify the Christian viewpoint as overwhelmingly negative. This was perhaps most famously argued in an influential article by Lynn White that traduced Christianity for being fundamentally anti-nature, obsessed with the metaphysical to the point of abrogating responsibility for the physical.[11] There has certainly been a deafening silence on the topic from the majority of Christian leaders historically, with some notable exceptions, like Francis of Assisi.[12] Even St Patrick – the Irish patron saint whose sometime mountain home I passed in Chapter 1 – referred to wolves more as metaphors for dodgy people than in and of themselves.[13] Yet the pages of the Bible are alive with biodiversity, in both literal and symbolic depictions. In recent time, Pope Francis's encyclical on 'care for creation' has further cemented the issue as one that should matter to the world's biggest religion.[14] And my own experience, of setting up the faith-based parent organisation of Jubilee Farm, where we worked with Christians of all stripes, as well as people of all backgrounds and beliefs, has cemented my belief that this particular meta-narrative can and should play an important role in stewarding our shared home, especially for those whose moral experiences it shapes.

However, as Britain and Ireland become both increasingly secular, on one hand, and multicultural, on the other, there are an increasing number of stories shaping our notions of right and wrong, and of ethics and morality. This includes how we relate to the world around us, including the potential reintroduction of big predators to our island homes. Paganism, for instance, is making a comeback.[15] Islam is growing rapidly, especially in England.[16] And an increasing number of people, especially younger generations, identify as atheist or agnostic. How will this diverse metaphysical mixture influence our debate? To be honest, I'm not sure it will, to any great degree. And if it does, it will probably be nuanced, as both our brief discussions of the pagan and Christian perspectives illustrate. My experience of researching religion and religiosity in relation to snow leopards in Nepal,[17] and of establishing Jubilee Farm, leads me to believe that there are often a wide range of stories and philosophies, religious and otherwise, by which we live our lives. Our uniquely personal moral experiences are a reflection of the diversity of meta-narratives that shape us all: past, present and future.

That said, there is a major philosophical fault line that I see developing in this debate, other than the one between those who wish to reintroduce large carnivores and those who don't, which we've explored at length. This is the clash between biocentrism and ecocentrism. Biocentrism focuses more on individual animals, and their welfare and rights, while ecocentrism's priority is populations of animals or even entire ecosystems.[18] The latter is often willing to sacrifice the welfare of certain individuals if it benefits the overall group. We touched on this debate in Chapter 8, when we considered the use of force and its contested application in Europe recently. We also briefly considered it in relation to hunting as a conservation tool in Chapter 9.

This matters in relation to this debate because, as we discussed in Chapter 11, reintroductions often suffer from high failure and mortality rates.[19] This raises ethical questions, such as what mortality levels for reintroduced lynx or wolves are acceptable. I could see a complex situation developing where animal rights groups weigh in on this. To add to this, they may even take issue with the extent to which farmers are safeguarding their livestock from these reintroduced predators. A bizarre and troubling court case to this effect was underway when I visited the Netherlands. An animal rights organisation took a sheep farmer to court, accusing him of not taking sufficient precautions to safeguard his sheep from wolves.[20] This was despite the fact that the farmer had multiple protection methods in place, including electric fencing. Thankfully, the judge threw out the case.

Our shared moral experiences, shaped by the stories we live by, are complex and nuanced. They combine with all these meta-narratives past and present, building up over time, like the stratigraphy of an ancient city such as Troy or Jericho, to create layered landscapes. These landscapes are both physical and metaphysical. They fuse our inner and outer worlds: our politics, philosophy, economics, ecology and psychology. This is the second concept that Martin introduced to me.

'In terms of the outworking of these narratives, they find expression in cultural landscapes. These have varied throughout history according to the culture of the time,' observed Martin.

I thought of Nepal and Malawi and Yellowstone, of their layers of nature and human nature. I thought of Britain and Ireland, and of their stratigraphy of pagan, Christian and secular stories, among others. I thought of Missoula, with its bears and students and joggers and pumpkins.

Martin continued: 'In practice, most landscapes exhibit layers of human history with different cultural landscapes built on top of each other. Rewilded landscapes are simply the new cultural landscapes of our time, reflecting what is important to society at this moment. But they do do something special compared to other cultural interventions, in the sense that they explicitly seek to break with the all-too-human perspective of the history of the land. Instead, they seek to restore a sense of narrative continuity with the deeper, older earth history.'

I remembered how, at the start of my journey, I initially felt farming and rewilding were on the same continuum of our engagement with nature. Just different ways of doing things to, with and for species other than our own. Based on visions and values that were sometimes different and sometimes similar. Now, nearing the end of my rewilding odyssey, I felt that my experiences and research reinforced this impression. But between the contrasting caricatures of wild and domestic across Britain and Ireland, and across the Netherlands, Switzerland and the USA, was a subjective spectrum of space that was a bit of both. On this continuum of wildness, where was wild enough for lynx and wolves and bears, and where wasn't?

Martin's parting thought was a powerful reminder of how all this philosophising could and would shape the future, whether it was wild or domestic or somewhere in between: 'Given that it is impossible to return to exact representations of past landscapes, moving forward by creating new landscapes is the only option.'

I nodded in agreement and, pertinently, Martin added, 'It's akin to writing a new chapter of a landscape that follows on from the preceding ones.'

Our philosophical bents, from moral experiences to layered landscapes, are a fulcrum of the debate about reintroducing lynx, wolves and bears to Britain and Ireland. Could and should these species fit into a new layer of our landscapes that is simultaneously vulnerable, resilient and functional? Despite their ecological underpinnings, rewilding and reintroducing remain fundamentally social processes underpinned by politics, philosophy, economics and psychology. The philosophy part gives us the vision and values, the ethics and morality, that contribute to a broad range of viewpoints on this contentious topic. For that reason, I still think that the philosophical case is a nuanced and subjective one. Cases plural, in fact, rather than case singular. There will be multiple moral experiences writing the next chapter in the story of British and Irish landscapes in the twenty-first century. We need to find ways to write this story together.

For that reason, I also think that the philosophical case that matters most is appreciating and understanding each other's meta-narratives and moral experiences, even if we don't always or often agree. We all share a layered landscape that is the ultimate form of common ground. For that, we don't just require broad minds but large hearts. Not just knowledge but compassion. In short, we need wisdom. Perhaps even the wisdom of Solomon, who, in the Christian story, was given it as a gift – 'the broadest of minds and the largest of hearts like the grains of sand upon the seashore'.[21]

In this complex and contested reintroduction debate, which will probably rumble on for the rest of the twenty-first century, may we prize wisdom above all else. For the love of wisdom, may we all seek broader minds and deeper hearts as we discuss and debate with each other across the layered landscapes we know and love. Wild, wild wisdom – like the grains of sand upon the seashore.

CHAPTER 15

Reconciliation

At the end of my journey, I went back to the beginning. On a cold winter's day, I drove east from my home in Ballymena, heading for the hills. Over a year had passed since I came to this place to begin my quest. The glorious Glens of Antrim stretched away to the north and south. Like the spine of the landscape, they gave structure to its soft edges, its flows and its eddies, its forests, fields and bogs. They called to me and I answered.

I told them of the vast turbine fields of Wyoming and of the cowboys who harness the wind. They showed me their windfarms, still pirouetting and glinting as sunshine broke through the cloudbank. I said I'd been with other mountains in Colorado, Montana and Switzerland: their beauty was not greater, just grander. The low winter sun picked out the pale palette of colours on the plateau around me: a tired yellow, a weary red, a weak russet. Their magnificence still slumbered under winter's spell, awaiting spring's embrace. The glens showed me their gentle grandeur, made all the more becoming because it was home. I admitted that I'd been back in that adopted home of mine across two seas; that the Dutch countryside did not have the air of abandonment and decline that characterised parts of this landscape. For amongst the farms and homes that were flourishing in the highlands of Co. Antrim were the skeletal remains of the many farms and homes that were gone forever. A gust of wind flung rain against the car. It ran down the windscreen like tears. These mountains mourned the dying decades of an epoch.

This place spoke to me. And I listened. It reached out its hand to me and I grasped it.

But I was not here to dissect the past. I was here to peer into the future. With one hand I did take hold of the hand of history. Yet with the other I seized tomorrow's outstretched palm. Through me, past, present and future fused.

Amidst the glorious Glens of Antrim, I arrived at Hugh's farm. We chewed the fat in the yard. The two pigs, or what was left of them, were now safely stowed in the freezer. I slipped on my wellies and we headed across the fields and up the hill.

In the year that had passed since I last visited Hugh, everything had changed. And in the timeless cycle of the seasons, everything was the same. The last batch of lambs, barely conceived on my last visit, were fattening nicely on the remaining grass. But these were really the last batch. Along with the ewes, in a small flock he'd spent years breeding, all Hugh's sheep were due to depart in the coming weeks. A pernicious shoulder injury had taken its toll on him. It was time to take a break from the heavy manual handling that shepherding involved.

'What are your plans for the farm?' I wondered.

Hugh had spoken before about how, since he had sold his remaining cattle, the grassland hadn't been quite the same.

'I think I'll rent the ground to a dairy farmer to run heifers over. The pasture will benefit from their coarse grazing and trampling and dunging. If I can allow this shoulder to heal, I'd be interested in getting a few Dexters or another small native breed in a few years.'

He continued with a mischievous smile: 'If the boss allows me, that is.'

Of all the livestock I'd encountered on my reintroduction research road trip, none were more vulnerable to predation than sheep, or their caprid cousins, goats. And while cattle and horses were occasionally vulnerable to wolves, especially where sheep and deer populations were lower, they were safe from the attentions of lynx, the most likely candidate for reintroduction, in Britain at least. As the uplands across both Britain and Ireland changed in the decades to come, would sheep depart and cattle take their place, as on Hugh's farm? And would this facilitate coexistence between farming and rewilding? It was one of several key questions that I was ruminating on.

I told Hugh about Shaun's vast sheep and cattle operation on the Wyoming/Utah border.

'That landscape-level transhumance is a dying art,' I lamented.

It had long since gone from these parts too. All that remained outside history were the ruins of the booley huts. On forgotten hillsides, only the wind whistled through these lonely monuments to a bygone era. Yet at Anderson Ranch in the Tom Miner Basin, and now soon on Hugh's land, the seasonal movement of cattle to capture the flush of summer grass lived on in a contemporary form. So too did the challenges of keeping livestock safe from predators. I talked about how Shaun trapped and shot coyotes, and translocated eagles. Closer to home, two sea eagles, visiting from Scotland, had been poisoned some months previously in the Antrim hills. Our conversation drifted to this topic, perhaps the closest proxy for large carnivore reintroductions in Britain and Ireland so far.

'From my conversations with farming organisations,' I ventured, 'sea eagles are held up as an example of the challenges of reintroductions, including applying deterrence, finance and force as coexistence tools. There's also the concern that the benefits from sea eagle tourism aren't being fairly allocated to those bearing the costs.'

Hugh had read about examples of farmers coming round to the idea in Scotland and in Kerry: 'I'd be happy to have them back in the Glens of Antrim – the occasional sea eagle wouldn't do as much damage as forty or fifty ravens.'

'And yet,' I quipped, 'perception is often reality: reintroducing animals with sharp claws and pointy teeth has a symbolic quality that exacerbates emotions, amongst those for and those against.'

We agreed perceptions could go either way, whether with sea eagles or with lynx. I shared my experience from Switzerland, where a lack of compromise over managing coexistence between livestock farming and a rapidly expanding wolf population had spiralled into a toxic and highly politicised conflict between people.

'That's the last thing we need in a place like this,' I added, as I gestured at the rolling hills around us.

From the high point on the farm, I could hear Hugh's neighbour whistling to his dog as he rounded up his sheep by quad. The temperature was falling and the wind was rising. We strolled back down the hill as we discussed electric vehicles and the changes underway in landscapes like these across these islands and the world, of which the actual or potential return of predators was but one factor among many. As I'd suspected on my first visit here, it was small-scale, often part-time and perpetually struggling farms like Hugh's that were at the centre of the debate about how the reintroduction process could be reconciled with agriculture. And as I'd seen, these farms and their families and communities were being buffeted by the winds of change.

This wind that whistles through these landscapes is a cold and unforgiving one. On it is borne the dying moans of an epoch: the beginning of the end of the industrial era. But on it is also borne the keening call of the future: the end of the beginning of the ecological era. The transition between them is not an easy one. The colours of this wind cover the full spectrum of human nature. As its full force is channelled by the brutal laws of nature, and by the inexorable march of time and technology, this wind rises to a shriek. Force becomes fury.

As we approach the denouement of our reintroduction drama, let us first look back over the key themes and characters and places that we have encountered along the way. For as we've seen throughout our story, it is by first looking backwards that we can best look forwards.

The colours of the wind

Psychology matters. It is within our inner landscapes that we make sense of the world around us, individually and collectively. By virtue of our evolutionary instincts and our cultural milieu, large carnivores roam these inner worlds too. We are conditioned to respond to them. Our powerful feelings towards predators are usually mixed and often polarised. The same goes for how we feel about their potential return to the outer landscapes of our island homes. Yet if our complex and multiplicitous minds allow us to hold multiple and even competing perspectives on this single issue within us, we can surely tolerate the same plurality of views between us.

History also matters. It helps us understand how and why 'where the wild things were' in Britain and Ireland became 'where the wild things weren't'. Bears and lynx vanished from our landscapes hundreds or thousands of years ago. Wolves have been absent for hundreds, with an especially strong shove from *Homo sapiens*. Island status accelerated this process and made it permanent. Until now. Like large carnivores themselves, the wild as a concept – and as a place – is complex and contested. Our perceptions of it, whether as utopia or dystopia, and large carnivores' symbolic representation thereof, are alive and well in the present.

History also has a firm grip on rewilding and reintroducing. These processes – the latter often as part of the former – drag the wild, kicking and screaming, from something distant in space and time to something proximate in present and future. Until now in Britain and Ireland, this has mostly involved birds of prey, and more recently, beavers and bison. But the calls to reintroduce lynx, sometimes wolves and occasionally bears are growing louder. A set of concrete plans for lynx reintroductions in England have already been quashed. Now more consultative processes about this in Scotland and England are well underway. Everybody may love a comeback kid on the sports field or in the movies. But not everybody loves a comeback carnivore, especially in their own backyard.

But are large carnivores the missing links necessary to reconcile our ecosystems? Of course we cannot fully replicate the ecological role of predators. But in performing a similar role, whether we harvest wild herbivores like deer through hunting, or domestic herbivores through farming, we are definitely apex predators in our own right. And coexistence provides us with the language and the tools to share landscapes with our fellow large carnivores and with our fellow humans, especially in the places, like the Dutch countryside, that lie midway on the spectrum of wildness, between urban metropolises and national parks like Yellowstone.

Farming with carnivores in these places is a challenge. Each deterrence tool we can employ to minimise the risks to livestock has pros and cons, whether it involves fences, guard animals, shepherds, corrals or collars. And all of them have costs: directly in terms of equipment and materials and labour; indirectly in terms of time spent on implementing these methods that is lost from some other important task.

In rural communities across Western Europe and North America, and perhaps one day in Britain and Ireland, these changes represent less minor amendments to daily routines and more systemic shifts of entire ways of life. Wolves especially, but also bears and to a lesser extent lynx, are lightning rods for this discontent.

It is not just landscapes that speak to us. Money talks. We listen. And when it comes to using financial tools, like compensation, insurance and proactive payments, to fund and facilitate coexistence between livestock farming and large carnivore conservation, we should listen most to those affected most. In theory, these financial payments transfer the economic burden of these livestock losses from rural farmers who generally don't want large carnivores around to urbanites who do. In practice, government bureaucracy, as well as the way money tends to alter human behaviour, make this approach tricky to get right, as vital as it is.

But the ethical complexities of financing coexistence pale next to the moral quandaries generated by using force as a management tool. Hazing, translocating and lethal control all have benefits and drawbacks. All of them should be employed with both careful thought and professional training. But all of them are absolutely necessary to deliver coexistence, especially in landscapes where space is shared between large carnivores, livestock farming and other human activities. What's more, when deterrence and finance fail, force gives agency to those living closest to carnivores: it makes them feel that their concerns have been heard and acted upon. It is an essential part of the deadly game between people and predators.

On top of these three coexistence tools, which try to minimise the risks of living with wolves, lynx, bears and the like, enterprise approaches seek to exploit the opportunities. They see potential over problems. They let our animals spirits roar. Through tourism particularly, but also via hunting and certification, enterprise methods also try to generate income from simply having these species around. As an entrepreneur, it's a topic I'm particularly passionate about. Aside from the usual array of barriers that lie in the way of turning bold ideas into successful businesses, there are some ethical and safety considerations. Never mind the fact that with cat species especially, most tourists will never see them. Yet the main challenge remains ensuring that the economic benefits from reintroductions flow to those paying their economic costs.

Arbitrating over all the uneven distributions of time, space and money in relation to coexisting with carnivores is the role of governance. Governance is not only the most important tool in our toolkit. It is also the foundation on which deterrence, finance, force and enterprise must rest. Governance provides vital mechanisms and procedures for people with wildly differing views on predators to disagree and yet still share space with other species and our own. It is here that the plurality of perspectives that we allow within us must be matched by the acceptance of the plurality of perspectives between us. Finding and firmly holding such common ground is essential.

Perhaps nature is the ultimate common ground. Perhaps it is the ultimate arbiter of our human nature. But it is certainly the primary argument behind the reintroduction of wolves, lynx and bears to Britain and Ireland, as well as the strongest element of the case for their return. The call of the wild is strong, as is the ecological case overall. Through the twin effects of trophic cascades and the landscape of fear, predators like these would have multiple benefits for our landscapes, especially through changing the number and behaviour of deer. Yet this would be less a simplistic 'wolves-change-rivers' version and more a case of predators as one factor among many affecting change in our ecosystems' dynamic equilibrium. Whether there is space and prey for large carnivores has been most widely modelled for lynx in both islands, followed by wolves, with very little attention paid to bears. There are also likely to be considerable challenges with ensuring effective reintroduction programmes that ensure long-term, genetically diverse populations of these species, in part because of our island geography.

Whether there is space and prey for large carnivores in Britain and Ireland is also highly subjective. What some see as landscapes that are wild enough for lynx, and maybe in parts of Scotland wolves, others see as places that are not wild enough any more. This subjectivity is reflected in the middling levels of support for predator reintroductions, with a probable split between rural and urban residents and livelihoods. We are political animals. And it is the political ecology of reintroductions – that these are shaped as much by power dynamics and economic factors as by ecology, if not more so – that leads me to conclude that the political case for the return of large carnivores is weak. The post-Brexit political atmosphere across our islands' two nations and five main political jurisdictions is febrile enough without adding pointy-toothed predators to the mix.

Economics is the third dimension of our political ecology perspective. And the economic case is fairly mixed. On one hand, the case for tourism, as well as for deer control, is strong. With the aforementioned caveat that most tourists will never see a lynx: it will be part of the 'brand' of the destination, much like the snow leopard in the shadow of Annapurna. In terms of the costs, the value of lost sheep is likely to be manageable for a well-designed financial compensation scheme. Yet the costs of training and equipping farmers to use deterrence methods are likely to far outstrip predation costs. Public funding, with its myriad pros and cons, will be required to fund this long term. The allocation of costs and benefits is also likely to be uneven and subjective within and between the different parts of Britain and Ireland. As is the value we attempt to put on intangible things, like the return of long-vanished species to wander the landscapes of our home sweet home once more.

Economics has its limits. To help us consider the intangible, it is to philosophy that we must turn. Here, as with economics, the case is mixed and subjective. For every point, there is a counter-point. For every perspective, a matching one weighed against it. Much of all this hinges on what and where wild is. And where, on the continuum of wildness, large carnivores belong. Our views on this are shaped by our collective moral expressions of nature as fragile, resilient and functional, as well as by the cultural landscapes that we share together. As with psychology, our inner and outer landscapes fuse in the philosophy of to reintroduce or not to reintroduce. And as with psychology, the most important reintroduction philosophy, I believe, is being open to the philosophies, moral expressions and cultural landscapes of others. In sharing these landscapes, whether with people or with predators, wisdom is required above all else: not only the broadest of minds but the largest of hearts; not only knowledge but compassion.

As a child of the 1990s, it would be remiss of me to finish this journey and this book without quoting a Disney princess, much as we began it that same way with Moana. In the words of Pocahontas: 'But if you walk the footsteps of a stranger / You'll learn things you never knew you never knew'. On my journey to write this book, I have walked in the footsteps of many strangers across large parts of the western hemisphere. I have learned many things I never knew I never knew. And I have tested and refined what I already believed.

That the colours of the wind that blow through the reintroduction landscape are many-hued. That they are the full spectrum of human nature. Channelled through the brutal laws of nature, and the inexorable march of time and technology, what do they mean for the potential return of wolves, lynx and bears to Britain and Ireland? Here, at the end of our rewilding odyssey, what can we conclude from all the people, places and predators that we've encountered together?

Coexisting in time and space

Seeking a final reconciliation of these questions and their answers, I headed north from Hugh's farm. Up and over the ridge to the next glen. Down the other side on a long, straight mountain road. I reached the car park at Dungonnell reservoir and parked up. At the very end of my journey, I had come back to the very beginning.

The temperature now hovered above freezing. A brief blast of sleet assaulted me as I stepped out of the car and put my walking boots on. I put my rain gear on for good measure: I thought I was in for a soaking. But the mood of the mercurial mountain weather shifted. As I set off around the road that skirts the dam, rays of weak winter sun again broke through the cloudbank. The surface of the water scintillated under their gaze, as squalls gusted across it. Around me the hills flowed together. It was lovely, as it always is.

My inner and outer landscapes collided. History and natural history came together. Past, present and future fused.

Here, where Ulster's last wolf roamed three centuries ago, was an appropriate place to think about the return of this and similar species to landscapes like these across Britain and Ireland. At the dawn of the industrial era, large carnivores finally vanished from our lands. Now, in the dying decades of this epoch, their potential return transfixes us.

For the last several centuries we have not had to think about living with predators in these isles. Trade-offs and disputes relating to coexisting with large carnivores are a process we comfortably outsourced to the Serengeti and David Attenborough. And even then, in the simplistic narratives of many wildlife documentaries, we usually saw only 'pristine' nature in 'pristine' parks, untroubled by having to share space with troublesome human beings. But much like the domestic energy and transport infrastructure required to reach net zero, and transition to an ecological era, these issues can no longer be off-shored. These debates force us to confront the messy reality of coexisting with nature and climate in ways that we'd previously been able to keep safely at arm's length: in the Middle East and North Sea with hydrocarbons, and in the Americas, Africa or Asia with apex predators.

As I walked past the eastern end of the dam and started to ascend onto the edge of the Garron plateau, I looked down into Glenariff. The steep sides of this Queen of the Glens I now knew intimately, from hauling my heavy rucksack up and down it over months of training for Nepal. I could see the Irish Sea beyond it, and further still, shining in a patch of distant sunshine, the mountains of Scotland. In the folds of these hills where the forests were, lynx reintroductions were being actively discussed. Far to the west, as I stared into the low winter light, I could make out the Sperrin mountains of mid Ulster, which also lay claim to hosting the last wolf in the province. Beyond that, hidden from sight, were the mountains of Donegal, where some suggested the species could return.

As I looked at this cross-section of my countries, I considered the complexities of their past and their present. As I looked almost from sea to shining sea, I thought about the challenges of their future. As I basked in the royal splendour of this greatest of glens, surrounded and sustained by the wonders of the world around me, everything became clear. All my thinking and research crystallised. My travels and knowledge and experiences reconciled. I came to my conclusions.

And the first of them is this: that at this moment in time, large carnivore reintroductions to Britain and Ireland, even of lynx, are a biological

Brexit. They are technically possible. But they are complex, contested and costly. Just like Brexit, they have an uneasy relationship with the past and with the future. Just like Brexit, they are emotive issues of vision, values and identity that get under our skin and into our psyche. Just like Brexit, there is a mismatch between those supporting them and those living with the consequences. And just like Brexit, the precise allocation of costs and benefits is not only subjective, but subject to a time lag before they become apparent.

Drawing this parallel at the culmination of this book is not an attempt to open a whole new Brexit can of worms here. We've already opened a large carnivore one of those. I acknowledge and respect the full spectrum of opinions on Brexit, and how this could never be captured by a binary vote alone. I also acknowledge and accept that the UK's membership of the EU, like the booley huts on the hillsides around me, is probably gone forever.

But I cannot deny that my experience of living close to the realities of Brexit in Northern Ireland has coloured my view of it. My childhood memories of criss-crossing a militarised border between Ireland North and South, and my desire to keep my children far from checkpoints with guns. My recent memories of buses burning in the streets near the school my children used to go to, as the post-Brexit reality and symbolism of a new border in the Irish Sea set in amidst Ulster's fragile, fractured communities. Haunted by the ghosts of empire, the Irish question has not only dominated British politics once again since 2016, but, for much of that period until the start of the Ukraine war in February 2022, European politics too. As perhaps the greatest political trauma of a generation, Brexit is too apt a comparison to ignore in our debate about returning predators to these very same places.

In particular, I look at all the political capital that has been squandered in this process instead of being applied to the most pressing issue, not only of our time, but of all time. This issue, as we keep coming back to again and again, is the reconciliation of the *oikos* – or 'home' in Greek – of ecology with the *oikos* of economy, so that the two are synchronised and not segregated. It is about transitioning from the industrial era to the ecological era, where our circular economies take, make and regenerate in symbiosis with the biosphere that sustains and inspires us all. This is arguably the most challenging task in human history, as well as the most pressing. Human society is the fulcrum on which it pivots. And a human society that is bogged down with the complexities of Brexit cannot apply itself fully to this momentous undertaking. Nor, I would argue, can a society bogged down with the complexities of reintroducing wolves, lynx and bears.

In other words, I think that the significant political capital and social conflict required to fulfil this particular rewilding vision will undermine the much bigger, more urgent and much more strategic goal of transitioning Britain and Ireland towards sustainability. Granted, some will argue these species could help contribute towards this goal by increasing carbon sequestration in upland forests that are less damaged by deer. Doubtless there is a case to be made on that front, and as something not really touched on in this book in any great detail, I welcome further research and debate on it.[1] But even so, I worry that the ecological benefits generated would not be enough to offset the depletion of political will that getting to that point would engender, and the knock-on effect this may have on other aspects of the process of ecological reconciliation. Put another way, the benefits of the landscape of fear ecologically may not be sufficient to counterbalance the costs of the landscape of fear socially. And like it or not, because of both evolution and culture, that social landscape of fear remains a significant determinant in this reintroduction debate: the idea that the big wolf really is bad; that animals with pointy teeth are to be feared; that grannies, children, pets and livestock all need to be locked up to protect them from the wild things that come in the night. With predators as with politics, perception remains reality.

In my own internal landscape, a significant fear of mine is that the British and Irish states will struggle with the complexities of managing and governing the reintroduction of large carnivores. This is informed not only by my experience of dealing with government as an entrepreneur doing something new and innovative. It is also based upon my experience of their handling of Brexit, especially the British government. As well as seeing the Dutch and Swiss states – relative paragons of efficiency and effectiveness – struggling with the natural recolonisation of wolves, never mind their reintroduction by humans.

In fact, a case could be made that the whole affair would be better off with governments of all sorts nowhere near it. But just like the

Half-Earth movement's desire for a people-free zone across half of the planet, that is not reality. It's la-la land. For better or for worse, the state will have to be an integral part of any reintroduction process, whether as arbiter, regulator or long-term funder of the suite of coexistence tools required to manage and govern this process.

I walked on through time and space. Around me the Garron plateau unfolded. I was keen to see if I could make it all the way across to look down on the next glen. But as I laboured over the damp and uneven ground and crested the ridge that I thought might give me the view I sought, I was disappointed. For the undulating sea of yellowed grass stretched on and on and on. It would take hours to cross it and then return to where I stood. But the light was starting to fade and I just didn't have time. A much longer journey lay ahead than I had anticipated.

The same is true when predators return to rural landscapes like these. In each of the places we've visited on our road trip – Switzerland, the Netherlands, Colorado, Wyoming and Montana – the debate about the return of large carnivores, whether people reintroduced them or through their natural recolonisation, took place over decades rather than years. Closer to home, we should also think in terms of decades rather than years. In fact, we are going to be arguing about this topic in Britain and Ireland for the rest of the twenty-first century. A long journey lies ahead of us.

Which brings me to my second main conclusion. I have been thinking about the complexities of large carnivore reintroductions to Britain and Ireland since about 2015. First, teaching a class on it at Queen's University Belfast in 2016. Then writing and presenting a short documentary on the topic for and with the Northern Ireland Science Festival in 2021. And most recently my 56 interviews and visits across seven countries for the research that underpins this book. The more I've immersed myself in the topic, the more I have come to believe that this is a difficult and complex idea. And especially with the notion of lynx reintroductions to Ireland, based on the flimsiest of archaeological findings, I am reminded of what Agatha Christie said: 'If the fact will not fit the theory – let the theory go.' As the evidence base currently stands, moving lynx to Ireland is not a reintroduction but a novel translocation, and therefore the creation of an entirely new ecosystem rather than the recreation of a previous one.

Paradoxically and simultaneously, I've come to believe that large carnivore reintroductions to our islands are more likely to occur in the coming decades than I first anticipated ten years ago. There are a number of key reasons for this. Firstly, social, and with it political, support may increase over time, as we become ever more ecologically aware on our progress towards an ever more ecological era. Secondly, technology is progressing rapidly. The combination of decreasing costs and sizes of sensors, coupled with the expansion of AI-powered models and systems able to collate and manage all this data in real time, may make coexistence with livestock farming more technically possible. It may allow better separation of farm animals and carnivores, even in shared landscapes, where the challenges of coexistence are most acute. But whether rewilders will accept sensors on and even in reintroduced predators – think less like the Serengeti and more like Jurassic Park – as a price worth paying for their return will be the subject of some debate.

Thirdly, and most significantly, systemic subsidy change, towards public money for public goods, will result in sweeping change across marginal and upland areas, as energy, forestry, conservation and recreation increasingly outcompete farming. In this ecological era, carbon cowboys who harness the wind and sequester carbon will rewrite the story of these landscapes, bringing opportunities as well as risks for farmers like Hugh. The ecological changes they usher in, effectively increasing habitat and prey for large carnivores through afforestation, will only increase calls for the return of lynx and wolves and maybe even bears, to complete and manage these new-yet-old ecosystems.

I began to head down from the plateau, retracing my steps. I passed two fellow hikers coming down from another part of the surrounding hills and nodded a greeting. More beams of sunlight broke through the clouds and shone on the water. They illuminated the path ahead, as the wind gusted around me.

The winds of change are blowing through these lands. They whistle through the epochal shifts from the industrial to the ecological era that will unfold in places like this. They howl through the debate about returning large carnivores to them. The colours of this wind are many-hued and human. For as much as ecology is used to justify reintroductions, it is politics, philosophy, economics and psychology that will dictate whether these ideas become reality. It is governance, more than any technical or even technological tool, that will dictate whether these ideas succeed or fail. It is human nature that shapes the reintroduction debate, and coexisting with carnivores, much more than nature does.

In turn the social processes that are rewilding and reintroducing, but also conserving and farming, require the tools found in the social sciences and humanities to understand and manage them better. They are mirrors into our souls. They illuminate our human nature. And they are needed not instead of the natural sciences but in harmony with them.

The sun was pouring through the split in the sky now. Golden light bathed this golden landscape. I basked in the glory. And the final piece of the puzzle slotted into place.

For farmers and rewilders alike, and for all of us on the reintroduction continuum, the intensity and complexity of emotions that we feel about large carnivores and their return is an indication of the depth of our love for the places we call home. It is also an indication of the depth of our ability to think and feel deeply. In our multiplicitous minds, if we can find room within us for a breadth of viewpoints on this one single issue, as I continue to after all my travels and research, then let us find room for this breadth of viewpoints between us as well.

This viewpoint may be coloured by whether you're reading this from the comfort of a leafy suburb, with a secure wage unmoored from the vagaries of the land, or from a small Scottish croft with a valuable and much-loved herd of pedigree rare-breed sheep. To the former, the idea of lynx or even wolf reintroductions may seem like a romantic ideal, a return to and of the wild. A landscape of wonder. To the latter, these very same ideas may appear as a mortal threat to their way of live and livelihood, a return to and of chaos. A landscape of fear.

Wherever you are on the rewilding and reintroducing spectrums, let us pause and put ourselves in the positions of both these people. Think for a minute from their perspective. Walk in the footsteps of these strangers. Know the spectrum of emotions and knowledge on this issue that they both possess. Know the layers of history that underpin all of this, as well as the stratigraphy of sentiment. Know things you never knew you never knew. Above all, let your inner and outer landscapes collide: the wonder and the fear; the chaos and control; the wild and the domestic.

As our inner and outer landscapes fuse in authentic connection we become fully human. Authentic connection to the land plus ourselves, others and whatever stories we tell to give meaning to it all: this makes us fully human. As well as with science, it is with stories – and their symbols and metaphors – that we make sense of the world around us: the hand of history here; the winds of change there. The reintroduction debate, about if and how we could learn to live with lynx, wolves and bears again in Britain and Ireland, is only getting started. Deeply rooted in time and space, this debate asks profound questions of our connection with ourselves, each other and the natural world.

Along the way, I hope that this particular story that you've read can help guide us on our voyage, within these beloved islands we call home. May it be one of reconciliation. May it lead to coexistence. With lynx and wolves and bears and, above all, with each other. For learning to live with these species is also about learning to live with our own.

Back on that windswept hillside, bathed in golden light, I journeyed on.

Notes

Chapter 1: The Lay of the Land

1. Williams, F. 2017. *The Nature Fix: Why Nature Makes Us Happier, Healthier and More Creative.* New York & London: W.W. Norton & Company.
2. PRONI. 2007. *Introduction to the Earl of Antrim Estate Papers.* Public Record Office of Northern Ireland.
3. GOAHS. 2020. 'Garron Tower'. https://antrimhistory.net/7110-2/
4. The Londonderry Arms Hotel. 2022. 'About Londonderry Arms'. https://londonderryarmshotel.com/about-londonderry-arms/
5. McKillop, F. 1996. *Glencloy: A Local History.* Belfast: Ulster Journals Ltd.
6. Hickey, K. 2013. *Wolves in Ireland: A Natural and Cultural History.* Dublin: Open Air.
7. Hetherington, D.A., Lord, T.C. and Jacobi, R.M. 2006. 'New evidence for the occurrence of Eurasian lynx (*Lynx lynx*) in medieval Britain', *Journal of Quaternary Science* 21.1: 1–108. https://doi.org/10.1002/jqs.960
8. Raye, L., 2017. The Eurasian lynx (*Lynx lynx*) in early modern Scotland. *Archives of Natural History* 44.2: 321–33. https://doi.org/10.3366/anh.2017.0452
9. Woodman, P., McCarthy, M. and Monaghan, N. 1997. 'The Irish quaternary fauna project', *Quaternary Science Reviews* 16.2: 129–59. https://doi.org/10.1016/S0277-3791(96)00037-6
10. Edwards, C.J. et al. 2011. 'Ancient hybridization and an Irish origin for the modern polar bear matriline', *Current Biology* 21: 1251–8. https://doi.org/10.1016/j.cub.2011.05.058
11. O'Regan, H.J. 2018. 'The presence of the brown bear *Ursus arctos* in Holocene Britain: a review of the evidence', *Mammal Review* 48.14: 229–44. https://doi.org/10.1111/mam.12127
12. Dowd, M. 2016. 'A remarkable cave discovery', *Archaeology Ireland* 30.2: 21–25. https://www.jstor.org/stable/43816774
13. Jepson, P. and Blythe, E. 2020. *Rewilding: The New Science of Ecological Recovery.* London: Icon.
14. Dennis, R. 2021. *Restoring the Wild: Rewilding Our Skies, Woods and Waterways.* London: William Collins.
15. Gow, D. 2020. *Bringing Back the Beaver: The Story of One Man's Quest to Rewild Britain's Waterways.* Vermont: Chelsea Green.
16. *The Guardian.* 2022. 'Wild bison return to UK for first time in thousands of years'. https://www.theguardian.com/environment/2022/jul/18/wild-bison-return-to-uk-for-first-time-in-thousands-of-years.
17. UK Government. 2018. 'Lynx reintroduction in Kielder Forest'. https://www.gov.uk/government/publications/lynx-reintroduction-in-kielder-forest

18. Bavin, D. and MacPherson, J. 2022. 'Lynx to Scotland: Assessing the social feasibility of returning Eurasian lynx to Scotland'. https://www.scotlandbigpicture.com/Images/2022/04/Lynx-to-Scotland-FINAL1.pdf. The Missing Lynx Project. 2024. 'About'. https://www.missinglynxproject.org.uk/about-project
19. https://dictionary.cambridge.org/dictionary/english/coexist

Chapter 2: Our Inner Landscapes

1. https://www.verywellmind.com/vicarious-trauma-the-cost-of-care-and-compassion-7377234
2. Jenkins, S.R. and Baird, S. 2002. 'Secondary traumatic stress and vicarious trauma: a validational study', *Journal of Traumatic Stress: Official Publication of the International Society for Traumatic Stress Studies* 15.5: 423–32. https://doi.org/10.1023/A:1020193526843
3. Hublin, J. et al. 2017. 'New fossils from Jebel Irhoud, Morocco and the pan-African origin of *Homo sapiens*', *Nature* 546: 289–92. https://doi.org/10.1038/nature22336
4. Hershkovits, I. et al. 2018. 'The earliest modern humans outside Africa', *Science* 359: 456–59. http://dx.doi.org/10.1126/science.aap8369
5. https://www.verywellmind.com/what-is-integenerational-trauma-5211898
6. Krippner, S. and Barrett, D. 2019. 'Transgenerational trauma: the role of epigenetics', *Journal of Mind and Behaviour* 40.1: 53–62.
7. Coelho, C.M. and Purkis, H. 2009. 'The origins of specific phobias: influential theories and current perspectives', *Review of General Psychology* 13.4: 335–48. https://doi.org/10.1037/a0017759
8. Hanson, J.H., Schutgens, M. and Leader-Williams, N. 2019. 'What factors best explain attitudes to snow leopards in the Nepal Himalayas?', *PLoS One* 14.10: p.e0223565. https://doi.org/10.1371/journal.pone.0223565; Hanson, J.H., Schutgens, M. and Baral, N. 2019. 'What explains tourists' support for snow leopard conservation in the Annapurna Conservation Area, Nepal?', *Human Dimensions of Wildlife* 24.1: 31–45. https://doi.org/10.1080/10871209.2019.1534293
9. McGilchrist, I. 2009. *The Master and His Emissary*. New Haven: Yale University Press.
10. Schwartz, R.C. 2021. *No Bad Parts: Healing Trauma and Restoring Wholeness with the Internal Family Systems Model*. Boulder: Sounds True.
11. https://societyforpsychotherapy.org/internal-family-systems-exploring-its-problematic-popularity/
12. Schwartz, R.C. 2021. *No Bad Parts: Healing Trauma and Restoring Wholeness with the Internal Family Systems Model*. Boulder: Sounds True.
13. *The Economist*. 2023. 'Bully by name'. 26 August.
14. Clayton, S. and Myers, G. 2015. *Conservation Psychology: Understanding and Promoting Human Care for Nature*. Chichester: John Wiley & Sons.
15. Selinske, M. et. al. 2018. 'Revisiting the promise of conservation psychology', *Conservation Biology* 32.6: 1464–68. https://doi.org/10.1111/cobi.13106

16. Carter, N.H. and Linnell, J.D.C. 2016. 'Co-adaptation is key to coexisting with large carnivores', *Trends in Ecology and Evolution* 31: 575–78. https://doi.org/10.1016/j.tree.2016.05.006; Ceaușu, S. et al. 2018. 'Governing trade-offs in ecosystem services and disservices to achieve human–wildlife coexistence', *Conservation Biology* 33: 543–53. https://doi.org/10.1111/cobi.13241
17. Carter, N.H., Baeza, A. and Magliocca, N.R. 2020. 'Emergent conservation outcomes of shared risk perception in human–wildlife systems', *Conservation Biology* 34: 903–14. https://doi.org/10.1111/cobi.13473; Rust, N. 2017. 'Can stakeholders agree on how to reduce human–carnivore conflict on Namibian livestock farms? A novel Q-methodology and Delphi exercise, *Oryx* 51.2: 339–46. https://doi.org/10.1017/S0030605315001179
18. Clayton, S. 2019. 'The psychology of rewilding', in Pettorelli, N. et al. (eds). *Rewilding*. Cambridge: Cambridge University Press, pp. 182–200.

Chapter 3: Where the Wild Things Aren't

1. https://www.np-zuidkennemerland.nl/299/national-park-zuidkennemerland
2. Redford, K.H. 2011. 'Misreading the conservation landscape', *Oryx* 45.3: 324–30. https://doi.org/10.1017/S0030605311000019; Ghosal, S. et. al. 2013. 'An ontological crisis? A review of large felid conservation in India', *Biodiversity and Conservation* 22.11: 2665–81. https://doi.org/10.1007/s10531-013-0549-6
3. Adams, W.B. 2013. *Against Extinction: The Story of Conservation*. London: Routledge.
4. Marwick, A. 2001. *The New Nature of History: Knowledge, Evidence, Language*. Basingstoke: Palgrave.
5. Caple, C. 2006. *Objects: Reluctant Witnesses to the Past*. London: Routledge.
6. Jepson, P. and Blythe, C. 2020. *Rewilding: The Radical New Science of Ecological Recovery*. London: Icon.
7. Adams, W.B. 2013. *Against Extinction: The Story of Conservation*. London: Routledge.
8. Melzer, A.M. 2014. 'The origin of the Counter-Enlightenment: Rousseau and the new religion of sincerity', *American Review of Political Science* 90.2: 344–60. https://doi.org/10.2307/2082889
9. Rousseau J.-J. 2004. *The Social Contract*. Translated by Cranston, M. London: Penguin Books.
10. Gilligan, D.S. 2006. *In the Years of the Mountains*. New York: Thunder's Mouth Press.
11. De Vos, J.M. et al. 2014. 'Estimating the normal background rate of species extinction', *Conservation Biology* 29.2: 452–62. https://doi.org/10.1111/cobi.12380
12. Jepson, P. and Blythe, C. 2020. *Rewilding: The Radical New Science of Ecological Recovery*. London: Icon.
13. Ponting, C. *A New Green History of the World: The Environment and the Collapse of Great Civilizations*. New York: Vintage.

14. Hallmann, C.A. et al. 2017. 'More than 75 percent decline over 27 years in total flying insect biomass in protected areas', *PLoS One* 12.10: e0185809. https://doi.org/10.1371/journal.pone.0185809
15. Fogarty, P. 2017. *Whittled Away: Ireland's Vanishing Nature*. Cork: The Collins Press; Monbiot, G. 2014. *Feral: Rewilding the Land, Sea and Human Life*. London: Penguin Books.
16. Barnett, R. 2019. *The Missing Lynx: The Past and Future of Britain's Lost Mammals*. London: Bloomsbury.
17. https://www.nma.gov.au/defining-moments/resources/extinction-of-thylacine
18. https://www.nhm.ac.uk/discover/news/2021/december/worlds-largest-eagle-hunted-unlike-any-other-bird-of-prey.html
19. Seidensticker, J., Christie, C. and Jackson, P. 1999. 'Tiger ecology: understanding and encouraging landscape patterns and conditions where tigers can persist', in Seidensticker, J., Christie, C. and Jackson, P. *Riding the Tiger: Tiger Conservation in Human-Dominated Landscapes*. Cambridge: Cambridge University Press.
20. MacArthur, R.H. and Wilson, E.O. 2001. *The Theory of Island Biogeography*. Vol. 1. Princeton: Princeton University Press.
21. Perry, R. 1978. *Wildlife in Britain and Ireland*. London: Croom Helm.
22. https://www.irishpost.com/life-style/which-us-presidents-have-irish-roots-193244
23. Edwards, C.J. et al. 2011. 'Ancient hybridization and an Irish origin for the modern polar bear matriline', *Current Biology* 21: 1251–58. https://doi.org/10.1016/j.cub.2011.05.058
24. O'Regan, H.J. 2018. 'The presence of the brown bear *Ursus arctos* in Holocene Britain: a review of the evidence', *Mammal Review* 48.4: 229–44. https://doi.org/10.1111/mam.12127
25. Woodman, P., McCarthy, M. and Monaghan, N. 1997. 'The Irish quaternary fauna project', *Quaternary Science Reviews* 16.2: 129–59. https://doi.org/10.1016/S0277-3791(96)00037-6
26. O'Donnell, L., 2018. 'Into the woods: revealing Ireland's Iron Age woodlands through archaeological charcoal analysis', *Environmental Archaeology* 23.3: 240–53. https://doi.org/10.1080/14614103.2017.1417094
27. Hetherington, D.A., Lord, T.C. and Jacobi, R.M. 2006. 'New evidence for the occurrence of Eurasian lynx (*Lynx lynx*) in medieval Britain', *Journal of Quaternary Science* 21.1: 1–108. https://doi.org/10.1002/jqs.960
28. Perry, R. 1978. *Wildlife in Britain and Ireland*. London: Croom Helm.
29. Raye, L., 2017. The Eurasian lynx (*Lynx lynx*) in early modern Scotland. *Archives of Natural History* 44.2: 321–33. https://doi.org/10.3366/anh.2017.0452
30. Ibid.
31. O'Connor, T. and Sykes, N., 2010. *Extinctions and Invasions: A Social History of British Fauna*. Oxford: Windgather Press.
32. Hickey, K. 2011. *Wolves in Ireland: A Natural and Cultural History*. Dublin: Four Courts Press.
33. Ibid.
34. Edwards, N. 1990. *The Archaeology of Early Medieval Ireland*. London & New York: Routledge.
35. Hickey, K. 2011. *Wolves in Ireland: A Natural and Cultural History*. Dublin: Four Courts Press.

36. Ibid.
37. Sands, D. 2022. 'Dewilding "Wolf-land"', *Conservation & Society* 20.3: 257–67. https://www.jstor.org/stable/27206636

Chapter 4: Comeback Kids

1. Aryal, A., Brunton, D. and Raubenheimer, D. 2013. 'Habitat assessment for the translocation of blue sheep to maintain a viable snow leopard population in the Mt Everest Region, Nepal', *Zoology and Ecology* 23.1: 66–82. https://doi.org/10.1080/21658005.2013.765634
2. Hanson, J.H. et al. 2020. 'Local attitudes to the proposed translocation of blue sheep *Pseudois nayaur* to Sagarmatha National Park, Nepal', *Oryx* 54.3: 344–50. https://doi.org/10.1017/S0030605318000157
3. Hanson, J.H. 2024. *Large Carnivore Reintroductions to Britain and Ireland: Farmers' Perspectives and Management Options*. Taunton: Nuffield Farming Scholarships Trust.
4. Hanson, J.H. et al. 2020. 'Local attitudes to the proposed translocation of blue sheep *Pseudois nayaur* to Sagarmatha National Park, Nepal', *Oryx* 54.3: 344–50. https://doi.org/10.1017/S0030605318000157
5. IUCN/SSC. 2013. *Guidelines for Reintroductions and Other Conservation Translocations*. Version 1.0. Gland: IUCN Species Survival Commission.
6. https://www.nirsg.com/red-kite/
7. Sands, D. 2023. *Living with the Wild: Rewilding Conflicts and Conservation Politics in Ireland*. PhD thesis. Oslo: Norwegian University of Life Sciences.
8. Soulé, M.E. and Noss, R. 1998. 'Rewilding and biodiversity: complementary goals for continental conservation', *Wild Earth* 8.3: 18–28.
9. Jørgensen, D. 2015. 'Rethinking rewilding', *Geoforum* 65: 482–88. https://doi.org/10.1016/j.geoforum.2014.11.01
10. The original video: Sustainable Human. 2014. 'How Wolves Change Rivers.' https://youtu.be/ysa5OBhXz-Q?si=LBO2T8Pg9Z1iT3RD; a recent critique: Beyers, R., et al. 2023. 'The Wolves of Yellowstone: Saviours of the Songbird or Pieces of the Puzzle?' in Convery, I. et al, *The Wolf: Culture, Nature, Heritage*. Martlesham & Rochester: Boydell & Brewer, pp. 249–258; in the European context: Gerber et al. 2024. 'Do recolonising wolves trigger non-consumptive effects in European ecosystems? A review of evidence', *Wildlife Biology* e01229. https://doi.org/10.1002/wlb3.01229
11. EC. 1992. 'Council Directive 92/43/EEC of 21 May 1992 on the conservation of natural habitats and of wild fauna and flora', *Official Journal of the European Communities* L 206, 7. https://eur-lex.europa.eu/legal-content/EN/TXT/PDF/?uri=CELEX:01992L0043-19971128
12. EC. 2012. *Status, Management and Distribution of Large Carnivores – Bear, Lynx, Wolf & Wolverine – in Europe*. Luxembourg: Office for Official Publications of the European Communities.
13. https://www.lcie.org/Largecarnivores/Brownbear.aspx
14. https://www.lcie.org/Largecarnivores/Wolf.aspx
15. https://www.lcie.org/Largecarnivores/Eurasianlynx.aspx
16. KORA Foundation. 2022. *50 Years of Lynx Presence in Switzerland*. KORA Report Nr 99e. Ittigen: KORA.

17. Garrote, G. et al. 2013. 'Human–felid conflict as a further handicap to the conservation of the critically endangered Iberian lynx', *European Journal of Wildlife Research* 59.2: 287–90. https://doi.org/10.1007/s10344-013-0695-x
18. Cracknell, S. 2021. *The Implausible Rewilding of the Pyrenees*. Self-published.
19. https://rewildingeurope.com/
20. Sandom, C.J. and Wynne-Jones, S. 2019. 'Rewilding a country: Britain as a case study', in Pettorelli, N., Durant, S.M. and du Toit, J.T. (eds). *Rewilding*. Cambridge: Cambridge University Press.
21. Henaughen, K. 2021. 'Time for action on sea eagles', *The Scottish Farmer*. https://www.thescottishfarmer.co.uk/news/19216569.time-action-sea-eagles/
22. Amos, I. 2022. 'Legal culls of "problem" beavers fall amid support for species expansion in Scotland, official figures show', *The Scotsman*. https://www.scotsman.com/news/environment/legal-culls-of-problem-beavers-fall-amid-support-for-species-expansion-in-scotland-official-figures-show-3705546
23. https://www.kentwildlifetrust.org.uk/projects/wilder-blean
24. https://www.gov.uk/government/publications/lynx-reintroduction-in-kielder-forest/lynx-reintroduction-in-kielder-forest-natural-england-advice-to-the-secretary-of-state#social-feasibility-and-risk-assessment
25. https://www.rewildingbritain.org.uk/about-us/what-we-say/policy/eurasian-lynx-position-statement
26. Murphy, K. et al. 2023. 'GIS-integrated agent-based simulations to model wolf reintroduction management scenarios in Ireland', *Authorea Preprint*. https://doi.org/10.22541/au.169070953.36409085/v1; Guilfoyle, C., Wilson-Parr, R. and O'Brien, J. 2023. 'Assessing the ecological suitability of the Irish landscape for the Eurasian lynx (*Lynx lynx*)', *Mammal Research* 68: 151–66. https://doi.org/10.1007/s13364-022-00670-2
27. https://www.bbc.co.uk/news/uk-northern-ireland-66134319

Chapter 5: Missing Links

1. https://www.nationaalparknieuwland.nl/en/the-park/oostvaardersplassen
2. https://www.theguardian.com/environment/2018/apr/27/dutch-rewilding-experiment-backfires-as-thousands-of-animals-starve
3. Beschta, R.L. and Ripple, W.J. 2009. 'Large predators and trophic cascades in terrestrial ecosystems of the western United States', *Biological Conservation* 142.11: 2401–14. https://doi.org/10.1016/j.biocon.2009.06.015; Hudson, P. et al. 2002. *The Ecology of Wildlife Diseases*. Oxford: Oxford University Press; Wan, X. et al. 2022. 'Broad-scale climate variation drives the dynamics of animal populations: a global multi-taxa analysis', *Biological Reviews* 97: 2174–94. https://doi.org/10.1111/brv.12888
4. Lennox, R. et al. 2022. 'The roles of humans and apex predators in sustaining ecosystem structure and function: contrast, complementarity and coexistence', *People and Nature* 4.5: 1071–82. https://doi.org/10.1002/pan3.10385
5. Blythe, C. and Jepson, P. 2020. *Rewilding: The Radical New Science of Ecological Recovery*. London: Icon.
6. Bonnot, N. et al. 2013. 'Habitat use under predation risk: hunting, roads and human dwellings influence the spatial behaviour of roe deer', *European*

Journal of Wildlife Research 59: 185–93. https://doi.org/10.1007/s10344-012-0665-8; Little, A.R. et al. 2016. 'Hunting intensity alters movement behaviour of white-tailed deer', *Basic and Applied Ecology* 17.4: 360–69. https://doi.org/10.1016/j.baae.2015.12.003

7. Guy, G. 2022. *Landscape Scale Management of Wild Deer to Combat Impacts on Habitats and Rural Enterprises and Produce Sustainable Venison in the UK*. Taunton: Nuffield Farming Scholarships Trust.
8. Bhatia, S. et al. 2020. 'Beyond conflict: exploring the spectrum of human–wildlife interactions and their underlying mechanisms', *Oryx* 54.5: 621–28. https://doi.org/10.1017/S003060531800159X
9. Hodgson, I. et al. 2020. *The State of Knowledge and Practice on Human–Wildlife Conflicts*. Geneva: Luc Hoffmann Institute.
10. Balmford, A. 2021. 'Concentrating vs. spreading our footprint: how to meet humanity's needs at least cost to nature', *Journal of Zoology* 315.2: 79–109. https://doi.org/10.1111/jzo.12920; Phalan, B. et al. 2011. 'Reconciling food production and biodiversity conservation: land sharing and land sparing compared', *Science* 333.6047: 1289–91. http://dx.doi.org/10.1126/science.1208742
11. Grass, I. et al. 2019. 'Land-sharing/-sparing connectivity landscapes for ecosystem services and biodiversity conservation', *People and Nature* 1.2: 262–72. https://doi.org/10.1002/pan3.21
12. Wilson, E.O. 2016. *Half-Earth: Our Planet's Fight for Life*. New York: Liveright.
13. Büscher, B. 2016. 'Half-Earth or Whole Earth? Radical ideas for conservation, and their implications', *Oryx* 51.3: 407–10. https://doi.org/10.1017/S0030605316001228

Chapter 6: The Lynx Will Lie Down with the Lamb

1. Dickman, A.J. et al. 2014. 'Carnivores, culture and "contagious conflict": multiple factors influence perceived problems with carnivores in Tanzania's Ruaha landscape', *Biological Conservation* 178: 19–27. https://doi.org/10.1016/j.biocon.2014.07.011; Jacobsen, K.S. et al. 'The importance of tangible and intangible factors in human–carnivore coexistence', *Conservation Biology* 35: 1233–44. https://doi.org/10.1111/cobi.13678
2. Carter, N.H., Baeza, A. and Magliocca, N. 2020. 'Emergent conservation outcomes of shared risk perception in human–wildlife systems', *Conservation Biology* 34: 903–14. https://doi.org/10.1111/cobi.13473
3. Hanson, J.H. 2022. 'Household conflicts with snow leopard conservation and impacts from snow leopards in the Everest and Annapurna regions of Nepal', *Environmental Management* 70.1: 105–16. https://doi.org/10.1007/s00267-022-01653-4
4. KORA Foundation. 2020. *25 Years of Wolf Presence in Switzerland: An Interim Assessment*.
Muri: KORA.
5. https://www.kora.ch/de/arten/wolf/uebergriffe-auf-nutztiere
6. Linnell, J.D.C., Odden, J. and Mertens, A. 2012. 'Mitigation methods for conflicts associated with carnivore depredation on livestock', in Boitani, L. and Powell, R.A. (eds). *Carnivore Ecology and Conservation: A Handbook of Techniques*. Oxford: Oxford University Press, pp. 314–32.

7. Van Eeden, L.M. et al. 2018. 'Carnivore conservation needs evidence-based livestock protection', *PLoS Biol* 16.9: e2005577. https://doi.org/10.1371/journal.pbio.2005577
8. Van Eeden, L.M. et al. 2018. 'Managing conflict between large carnivores and livestock', *Conservation Biology* 32: 26–34. https://doi.org/10.1111/cobi.12959
9. https://www.agridea.ch/en/
10. Vogt, K. et al. 2022. *Wirksamkeit von Herdenschutzmassnahmen und Wolfsabschüssen unter Berücksichtigung räumlicher und biologischer Faktoren*. Bericht in Zusammenarbeit mit AGRIDEA. KORA Bericht Nr 105. Muri bei Bern: KORA. https://www.kora.ch/?action=get_file&id=154&resource_link_id=2a6
11. Moreira-Arce, D. et al. 2018. 'Management tools to reduce carnivore-livestock conflicts: current gap and future challenges', *Rangeland Ecology & Management* 71.3: 389–94. https://doi.org/10.1016/j.rama.2018.02.005
12. Iliopoulos, Y. et al. 2019. 'Tools for co-existence: fladry corrals efficiently repel wild wolves (*Canis lupus*) from experimental baiting sites', *Wildlife Research* 46.6: 484–98. https://doi.org/10.1071/WR18146
13. Jackson, R., Wangchuk, R. and Hillard, D. 2001. *Grassroots Measures to Protect the Endangered Snow Leopard from Herder Retribution: Lessons Learned from Predator-Proofing Corrals in Ladakh*. Los Gatos, CA: The Snow Leopard Conservancy. https://snowleopardconservancy.org/pdf/Jackson%20et%20al%20corral%20predproof%20paper.pdf
14. Khorozyan, I. et al. 2020. 'Studded leather collars are very effective in protecting cattle from leopard (*Panthera pardus*) attacks', *Ecological Solutions and Evidence* 1.1: e12013. https://doi.org/10.1002/2688-8319.12013
15. Whitehouse-Tedd, K. et al. 2020. 'Reported livestock guarding dog–wildlife interactions: implications for conservation and animal welfare', *Biological Conservation* 241: 108249. https://doi.org/10.1016/j.biocon.2019.108249; Smith, B.R. et al. 2020. 'The ecological effects of livestock guarding dogs (LGDs) on target and non-target wildlife', *Journal of Vertebrate Biology* 69.3: 20103, 1–17. https://doi.org/10.25225/jvb.20103
16. https://www.aramis.admin.ch/Texte/?ProjectID=49462&Sprache=en-US
17. https://pasturs-voluntaris.ch/

Chapter 7: Money Talks

1. Hanson, J.H. 2022. 'Household conflicts with snow leopard conservation and impacts from snow leopards in the Everest and Annapurna regions of Nepal', *Environmental Management* 70.1: 105–16. https://doi.org/10.1007/s00267-022-01653-4
2. Chen, P. et al. 2016. 'Human–carnivore coexistence in Qomolangma (Mt. Everest) nature reserve, China: patterns and compensation', *Biological Conservation* 197: 18–26. https://doi.org/10.1016/j.biocon.2016.02.026
3. https://snowleopardresearchnepal.wordpress.com/2014/02/19/the-snow-leopard-bank-ltd/#more-67
4. Ravenelle, J. and Nyhus, P.J. 2017. 'Global patterns and trends in human–wildlife conflict compensation', *Conservation Biology* 31: 1247–56. https://doi.org/10.1111/cobi.12948
5. Interprovincial Overleg. 2023. *Addendum Interprovinciaal wolvenplan*. Den Haag: IPO. https://www.bij12.nl/onderwerp/wolf/beleid-en-organisatie/#provinciaal-wolvenbeleid-en-provinciale-wolvencommissies

6. https://www.swissinfo.ch/eng/society/protecting-livestock-from-predators-costs-swiss-taxpayers-millions/46345714
7. https://www.bafu.admin.ch/bafu/de/home/themen/biodiversitaet/mitteilungen.msg-id-94148.html
8. https://publicaties.bij12.nl/maak-kennis-met-bij12/unit-faunazaken
9. https://agriculture.ec.europa.eu/common-agricultural-policy/financing-cap/cap-funds_en
10. CPW. 2023. *Colorado Wolf Restoration and Management Plan*. Denver: Colorado Parks and Wildlife.
11. Ibid.
12. Milind, W. et al. 2016. 'A theoretical model of community operated compensation scheme for crop damage by wild herbivores', *Global Ecology and Conservation* 5: 58–70. https://doi.org/10.1016/j.gecco.2015.11.012
13. Alexander, J.S. et al. 2021. 'Assessing the effectiveness of a community-based livestock insurance program', *Environmental Management* 68: 87–99. https://doi.org/10.1007/s00267-021-01469-8
14. Zabel, A. and Holm-Müller, K. 2008. 'Conservation performance payments for carnivore conservation in Sweden', *Conservation Biology* 22.2: 247–51. https://www.jstor.org/stable/20183378
15. Dickman, A.J., MacDonald, E.A. and Macdonald, D.W. 2011. 'A review of financial instruments to pay for predator conservation and encourage human–carnivore coexistence', *PNAS* 138.10: 13937–44. https://doi.org/10.1073/pnas.1012972108
16. https://sites.warnercnr.colostate.edu/centerforhumancarnivorecoexistence/wolf-conflict-reduction-fund/
17. Bautista, C. et al. 2019. 'Large carnivore damage in Europe: analysis of compensation and prevention programs', *Biological Conservation* 235: 308–16. https://doi.org/10.1016/j.biocon.2019.04.019
18. Bulte, E.H. and Rondeau, D. 2005. 'Why compensating wildlife damages may be bad for wildlife', *Journal of Wildlife Management* 69: 14–19. https://www.jstor.org/stable/3803581

Chapter 8: A Deadly Game

1. Canadian lynx (*Lynx canadensis*) being a close cousin of, and similar size to, the Eurasian lynx (*Lynx lynx*).
2. Hodgson, I.D. et al. 2020. *The State of Knowledge and Practice on Human–Wildlife Conflicts*. Gland, Switzerland: The Luc Hoffman Institute.
3. Binley, A. 2022. 'Paintballs to be shot at Dutch wolves in bid to make them less tame', *BBC Europe*. https://www.bbc.co.uk/news/world-europe-63499054
4. Goodrich, J.M. et al. 2011. 'Conflicts between Amur (Siberian) tigers and humans in the Russian Far East', *Biological Conservation* 144.1: 584–92. https://doi.org/10.1016/j.biocon.2010.10.016
5. Petracca, L.S. et al. 2019. 'The effectiveness of hazing African lions as a conflict mitigation tool: implications for carnivore management', *Ecosphere* 10.12: e02967. https://doi.org/10.1002/ecs2.2967
6. https://www.nature.scot/professional-advice/protected-areas-and-species/protected-species/reintroducing-native-species/scottish-code-conservation-translocations

7. Thomas, S. et al. 2023. 'Evaluating the performance of conservation translocations in large carnivores across the world', *Biological Conservation* 279: 109909. https://doi.org/10.1016/j.biocon.2023.109909
8. Massei, G. et al. 2010. 'Can translocations be used to mitigate human–wildlife conflicts?', *Wildlife Research* 37: 428–39. https://doi.org/10.1071/WR08179
9. Fontúrbel, F.E. and Simonetti, J.A. 2011. 'Translocations and human–carnivore conflicts: problem solving or problem creating?', *Wildlife Biology* 17: 217–24. https://doi.org/10.2981/10-091
10. https://www.politico.eu/article/fate-wolf-killed-pony-european-commission-ursula-von-der-leyen/
11. https://www.swissinfo.ch/eng/sci-tech/when-it-s-legal-to-shoot-the-wrong-wolf/48264920
12. https://www.reuters.com/article/swiss-wolves-idINKCN26G0WO/
13. https://www.politico.eu/article/fate-wolf-killed-pony-european-commission-ursula-von-der-leyen/
14. https://www.wired.com/story/rewilding-italy-bear-attack/
15. https://www.bbc.com/news/world-europe-65720066
16. https://www.euronews.com/my-europe/2023/09/13/eu-parliament-divided-on-plans-to-review-wolves-conservation-status; https://www.rte.ie/news/world/2024/0925/1471837-eu-wolf-protection-status/
17. https://www.swissinfo.ch/eng/politics/swiss-wolf-culls-suspended-until-court-reaches-verdict/49100592
18. Moreira-Arce, D. et al. 2018. 'Management tools to reduce carnivore–livestock conflicts: current gap and future challenges', *Rangeland Ecology & Management* 71.3: 389–94. https://doi.org/10.1016/j.rama.2018.02.005
19. Van Eeden, L.M. et al. 2018. 'Managing conflict between large carnivores and livestock', *Conservation Biology* 32: 26–34. https://doi.org/10.1111/cobi.12959
20. McManus, J. et al. 2015. 'Dead or alive? Comparing costs and benefits of lethal and non-lethal human–wildlife conflict mitigation on livestock farms', *Oryx* 49.4: 687–95. https://doi.org/10.1017/S0030605313001610
21. Cassidy, K. et al. 2023. 'Human-caused mortality triggers pack instability in gray wolves', *Frontiers in Ecology and the Environment* 21.8: 356–62. https://doi.org/10.1002/fee.2597
22. Monbiot, G. 2014. *Feral: Rewilding the Land, Sea and Human Life*. London: Penguin.

Chapter 9: Animal Spirits

1. Reynolds, P.C. and Braithwaite, D. 2001. 'Towards a conceptual framework for wildlife tourism', *Tourism Management* 22.1: 31–42. https://doi.org/10.1016/S0261-5177(00)00018-2
2. Duffield, J. 2019. 'Shopping for wolves: using nonmarket valuation for informing conservation districts', in Rasker, R. (ed.) *People and Public Lands*. Bozeman: Headwaters Economics.
3. https://mountainjournal.org/here-is-a-bearest-economy-most-american-states-would-die-to-have
4. Schutgens, M., Hanson, J.H. and Baral, N. 2018. 'Visitors' willingness to pay for snow leopard *Panthera uncia* conservation in the Annapurna

Conservation Area, Nepal', *Oryx* 53.4: 633–42. https://doi.org/10.1017/S0030605317001636
5. Reynolds, P.C. and Braithwaite, D. 2001. 'Towards a conceptual framework for wildlife tourism', *Tourism Management* 22.1: 31–42. https://doi.org/10.1016/S0261-5177(00)00018-2
6. Adam, W. 2004. *Against Extinction: The Story of Conservation*. London: Earthscan.
7. Treves, A., Santiago-Ávila, F.J. and Putrevu, K. 2019. 'Quantifying the effects of delisting wolves after the first state began lethal management', *European Journal of Wildlife Research* 65: 87. https://doi.org/10.7717/peerj.11666
8. https://mountainjournal.org/wyoming-aspires-to-make-money-selling-grizzly-tags
9. Treves, A. and Jones, S.M. 2010. 'Strategic tradeoffs for wildlife-friendly eco-labels', *Frontiers in Ecology and the Environment* 8: 491–8. https://doi.org/10.1890/080173
10. https://wildlifefriendly.org/

Chapter 10: Common Ground

1. https://ourcity.fcgov.com/rooted-in-community#:~:text=The%20City%20of%20Fort%20Collins,frequent%20and%20sound%20management%20practices
2. https://dictionary.cambridge.org/dictionary/english/governance
3. https://dictionary.cambridge.org/dictionary/english/governance
4. Hardin, G. 1968. 'The Tragedy of the Commons', *Science* 162: 1243–48. https://doi.org/10.1126/science.162.3859.1243
5. Ostrom, E. 1990. *Governing the Commons*. Cambridge: Cambridge University Press.
6. Bar-Tal, D., Halperin, E. and De Rivera, J. 2007. 'Collective emotions in conflict situations: societal implications', *Journal of Social Issues* 63.2: 441–60. https://doi.org/10.1111/j.1540-4560.2007.00518.x
7. Hodgson, I. et al. 2020. *The State of Knowledge and Practice on Human–Wildlife Conflicts*. Gland: The Luc Hoffman Institute.
8. Madden, F. and McQuinn, B. 2014. 'Conservation's blind spot: the case for conflict transformation in wildlife conservation', *Biological Conservation* 178: 97–106. https://doi.org/10.1016/j.biocon.2014.07.015
9. Marchini, S. et al. 2021. 'Planning for human–wildlife coexistence: conceptual framework, workshop process, and a model for transdisciplinary collaboration', *Frontiers in Conservation Science* 2. https://doi.org/10.3389/fcosc.2021.752953
10. Lute, M.L. et al. 2020. 'Conservation professionals' views on governing for coexistence with large carnivores'. *Biological Conservation* 248: 108668. https://doi.org/10.1016/j.biocon.2020.108668
11. NatureScot. 2020. *White-Tailed Eagle Action Plan Extension 2021–2024*. https://www.nature.scot/doc/white-tailed-eagle-action-plan-extension-2021-2024#1.+Introduction
12. Auster, R.E., Barr, S.W. and Brazier, R.E. 2022. 'Renewed coexistence: learning from steering group stakeholders on a beaver reintroduction

project in England', *European Journal of Wildlife Research* 68: 1. https://doi.org/10.1007/s10344-021-01555-6
13. https://www.nature.scot/professional-advice/protected-areas-and-species/protected-species/reintroducing-native-species/scottish-code-conservation-translocations
14. https://www.gov.uk/government/publications/reintroductions-and-conservation-translocations-in-england-code-guidance-and-forms

Chapter 11: The Call of the Wild

1. Maslow, A.H. 1943. 'Preface to motivation theory', *Psychosomatic medicine* 5.1: 85–92. https://doi.org/10.1097/00006842-194301000-00012
2. Jepson, P. and Blythe, C. 2020. *Rewilding: The Radical New Science of Ecological Recovery*. London: Icon.
3. https://www.theguardian.com/commentisfree/2023/aug/11/britain-deer-population-ecological-disaster-wolves-humans-predators
4. https://www.irishexaminer.com/lifestyle/outdoors/arid-41205349.html
5. Alston, J.M. et al. 2019. 'Reciprocity in restoration ecology: when might large carnivore reintroduction restore ecosystems?', *Biological Conservation* 234: 82–89. https://doi.org/10.1016/j.biocon.2019.03.021
6. Stepkovitch, B., Kingsford, R.T. and Moseby, K.E. 2022. 'A comprehensive review of mammalian carnivore translocations', *Mammalian Review* 52: 554–72. https://doi.org/10.1111/mam.12304
7. Gerber et al. 2023. 'Do recolonising wolves trigger non-consumptive effects in European ecosystems? A review of evidence', *Authorea Preprints*. https://doi.org/10.22541/au.169625166.69810849/v1
8. Clark-Wolf, T.J. and Hebblewhite, M. 2022. 'Trophic cascades as a basis for rewilding', in Hawkins, S. et al (eds) *Routledge Handbook of Rewilding*. London: Routledge, pp. 57–67; Hobbs, N.T. et al. 2024. 'Does restoring apex predators to food webs restore ecosystems? Large carnivores in Yellowstone as a model system', *Ecological Monographs* 94.2: e1598. https://doi.org/10.1002/ecm.1598
9. Brice, E.M., Larsen, E.J. and MacNulty, D.R. 2022. 'Sampling bias exaggerates a textbook example of a trophic cascade', *Ecology Letters* 25.1: 177–88. https://doi.org/10.1111/ele.13915
10. Guilfoyle, C., Wilson-Parr, R. and O'Brien, J. 2023. 'Assessing the ecological suitability of the Irish landscape for the Eurasian lynx (*Lynx lynx*)', *Mammal Research* 68: 151–66. https://doi.org/10.1007/s13364-022-00670-2
11. Hetherington, D.A. 2005. *The Feasibility of Reintroducing the Eurasian lynx Lynx lynx to Scotland*. PhD thesis. Aberdeen: University of Aberdeen; Hetherington, D.A. and Gorman, M.L. 2007. 'Using prey densities to estimate the potential size of reintroduced populations of Eurasian lynx', *Biological Conservation* 137: 37–44. https://doi.org/10.1016/j.biocon.2007.01.009
12. Johnson, R. and Greenwood, S. 2020. 'Assessing the ecological feasibility of reintroducing the Eurasian lynx (*Lynx lynx*) to southern Scotland, England and Wales', *Biodiversity Conservation* 29: 771–97. https://doi.org/10.1007/s10531-019-01909-2
13. Murphy, K. et al. 2023. 'GIS-integrated agent-based simulations to model wolf reintroduction management scenarios in Ireland', *Authorea Prepints*. 30 July. https://doi.org/10.22541/au.169070953.36409085/v1

14. Gwynn, V. and Symeonakis, E. 2022. 'Rule-based habitat suitability modelling for the reintroduction of the grey wolf (*Canis lupus*) in Scotland', *PLoS One* 17.10: e0265293. https://doi.org/10.1371/journal.pone.0265293
15. Ordiz, A. et al. 2020. 'Wolf habitat selection when sympatric or allopatric with brown bears in Scandinavia', *Scientific Reports* 10: 9941. https://doi.org/10.1038/s41598-020-66626-1
16. Carden, R.F., Carlin, C.M., Marnell, F., Mcelholm, D., Hetherington, J. and Gammell, M.P., 2011. Distribution and range expansion of deer in Ireland. *Mammal Review* 41.4: 313–25.
17. Guilfoyle, C., Wilson-Parr, R. and O'Brien, J. 2023. 'Assessing the ecological suitability of the Irish landscape for the Eurasian lynx (*Lynx lynx*)', *Mammal Research* 68: 151–66. https://doi.org/10.1007/s13364-022-00670-2
18. Twining, J.P. et al. 2022. 'Restoring vertebrate predator populations can provide landscape-scale biological control of established invasive vertebrates: insights from pine marten recovery in Europe', *Global Change Biology* 28: 5368–84. https://doi.org/10.1111/gcb.16236
19. http://irishdeercommission.ie/
20. Murphy, K. et al. 2023. 'GIS-integrated agent-based simulations to model wolf reintroduction management scenarios in Ireland', *Authorea Preprints*. 30 July. https://doi.org/10.22541/au.169070953.36409085/v1
21. https://bds.org.uk/information-advice/about-deer/deer-species/
22. Hetherington, D.A. and Gorman, M.L. 2007. 'Using prey densities to estimate the potential size of reintroduced populations of Eurasian lynx', *Biological Conservation* 137: 37–44. https://doi.org/10.1016/j.biocon.2007.01.009
23. Gwynn, V. and Symeonakis, E. 2022. 'Rule-based habitat suitability modelling for the reintroduction of the grey wolf (*Canis lupus*) in Scotland', *PLoS One* 17.10: e0265293. https://doi.org/10.1371/journal.pone.0265293
24. Alston, J.M. et al. 2019. 'Reciprocity in restoration ecology: when might large carnivore reintroduction restore ecosystems?', *Biological Conservation* 234: 82–89. https://doi.org/10.1016/j.biocon.2019.03.021
25. Twining, J.P. et al. 2022. 'Restoring vertebrate predator populations can provide landscape-scale biological control of established invasive vertebrates: insights from pine marten recovery in Europe', *Global Change Biology* 28: 5368–84. https://doi.org/10.1111/gcb.16236
26. Hetherington, D.A. and Gorman, M.L. 2007. 'Using prey densities to estimate the potential size of reintroduced populations of Eurasian lynx', *Biological Conservation* 137: 37–44. https://doi.org/10.1016/j.biocon.2007.01.009
27. Burak, M.K. et al. 2023. 'Context matters when rewilding for climate change', *People and Nature* 6.2: 507–18. https://doi.org/10.1002/pan3.10609
28. See, for example, Selonen, V. et al. 2022. 'Invasive species control with apex predators: increasing presence of wolves is associated with reduced occurrence of the alien raccoon dog', *Biological Invasions* 24: 3461–74. https://doi.org/10.1007/s10530-022-02850-2; Roos, S. et al. 2018. 'A review of predation as a limiting factor for bird populations in mesopredator-rich landscapes: a case study of the UK', *Biological Reviews* 93.4: 1915–37. https://doi.org/10.1111/brv.12426; Elmhagen, B. et al. 2010. 'Top predators, mesopredators and their prey: interference ecosystems along bioclimatic productivity gradients', *Journal of Animal Ecology* 79.4: 785–94. https://doi.org/10.1111/j.1365-2656.2010.01678.x

29. Hayhow, D.B. et al. 2019. *State of Nature 2019*. National Biodiversity Network. https://nbn.org.uk/wp-content/uploads/2019/09/State-of-Nature-2019-UK-full-report.pdf; National Parks and Wildlife Service. 2019. *The Status of EU Protected Habitats and Species in Ireland*. https://www.npws.ie/sites/default/files/publications/pdf/NPWS_2019_Vol1_Summary_Article17.pdf
30. Jepson, P. and Blythe, C. 2020. *Rewilding: The Radical New Science of Ecological Recovery*. London: Icon.
31. Stepkovitch, B., Kingsford, R.T. and Moseby, K.E. 2022. 'A comprehensive review of mammalian carnivore translocations', *Mammalian Review* 52: 554–72. https://doi.org/10.1111/mam.12304
32. Thomas, S. et al. 2023. 'Evaluating the performance of conservation translocations in large carnivores across the world', *Biological Conservation* 279: 109909. https://doi.org/10.1016/j.biocon.2023.109909
33. MacArthur, R.H. and Wilson, E.O. 2001. *The Theory of Island Biogeography*. Vol. 1. Princeton: Princeton University Press.

Chapter 12: Political Animals

1. Platt., S. 1993. *Respectfully Quoted: A Dictionary of Quotations*. London: Barnes & Noble.
2. Nolan, C. 2004. *Ethics and Statecraft: The Moral Dimension of International Affairs*. Westport: Greenwood.
3. Robbins, P. 2019. *Political Ecology: A Critical Introduction*. 3rd edition. London: Wiley.
4. Sands, D. 2022. 'Dewilding "Wolf-land"', *Conservation & Society* 20.3: 257–67. https://www.jstor.org/stable/27206636
5. Pheby, C. 2020. 'Which animals do Britons support reintroducing to the UK? YouGov survey results of 2083 adults in GB'. https://yougov.co.uk/topics/science/articles-reports/2020/01/28/third-brits-would-reintroduce-wolves-and-lynxes-uk
6. Research by the Macaulay Land Use Research Institute, cited in Watson Featherstone, A. 1997. 'The wild heart of the Highlands', *Ecos* 18: 48–61.
7. https://www.rewildingbritain.org.uk/press-hub/lynx-and-beaver-reintroductions-should-be-part-of-any-green-agreement-with-snp-says-coalition
8. Smith, D. et al. 2016. *Reintroduction of the Eurasian Lynx to the United Kingdom: Results of a Public Survey*. Lynx UK Trust. http://insight.cumbria.ac.uk/id/eprint/3188/1/lynxinterimsurvey.pdf
9. Hawkins, S.A. et al. 2020. 'Community perspectives on the reintroduction of Eurasian lynx (*Lynx lynx*) to the UK', *Restoration Ecology* 28: 1408–18. https://doi.org/10.1111/rec.13243
10. Bavin, D. et al. 2023. 'Stakeholder perspectives on the prospect of lynx *Lynx lynx* reintroduction in Scotland', *People and Nature* 5: 950–67. https://doi.org/10.1002/pan3.10465
11. Research by the Macaulay Land Use Research Institute, cited in Watson Featherstone, A. 1997. 'The wild heart of the Highlands', *Ecos* 18: 48–61.
12. Smith, D. et al. 2016. *Reintroduction of the Eurasian Lynx to the United Kingdom: Results of a Public Survey*. Lynx UK Trust. http://insight.cumbria.ac.uk/id/eprint/3188/1/lynxinterimsurvey.pdf

13. Wilson, S. and Campera, M. 2024. 'The perspectives of key stakeholders on the reintroduction of apex predators to the United Kingdom', *Ecologies* 5.1: 52–67. https://doi.org/10.3390/ecologies5010004
14. Tan, C.K.W., Shepherd-Cross, J. and Jacobsen, K.S. 2024. 'Farmers' attitudes and potential culling behaviour on the reintroduction of lynx to the UK', *European Journal of Wildlife Research* 70.1: 3. https://doi.org/10.1007/s10344-023-01751-6
15. Hanson, J. 2024. *Large Carnivore Reintroductions to Britain and Ireland: Farmers' Perspectives and Management Options*. Taunton: Nuffield Farming Scholarships Trust.
16. Dressel, S., Sandström, C. and Ericsson, G. 2015. 'A meta-analysis of studies on attitudes toward bears and wolves across Europe 1976–2012', *Conservation Biology* 29: 565–74. https://doi.org/10.1111/cobi.12420
17. Behr, D.M., Ozgul, A. and Cozzi, G. 2017. 'Combining human acceptance and habitat suitability in a unified socio-ecological suitability model: a case study of the wolf in Switzerland', *Journal of Applied Ecology* 54: 1919–29. https://doi.org/10.1111/1365-2664.12880
18. Convery, I. et al. 2023. 'The case for wolves in the UK', in Convery, I. et al. (eds). *The Wolf: Culture. Nature. Heritage*. Woodbridge & Rochester: Boydell Press, pp. 295–315.
19. HM Government. 2018. *A Green Future: Our 25 Year Plan to Improve the Environment*. https://www.gov.uk/government/publications/25-year-environment-plan
20. Government of Ireland. 2024. *Ireland's 4th National Biodiversity Action Plan*. https://www.gov.ie/en/publication/93973-irelands-4th-national-biodiversity-action-plan-20232030/
21. https://www.rewildingbritain.org.uk/press-hub/lynx-and-beaver-reintroductions-should-be-part-of-any-green-agreement-with-snp-says-coalition

Chapter 13: Home Sweet Home

1. https://wildlifefriendly.org/
2. See our research article for all the data and references relevant to this paragraph: Schutgens, M., Hanson, J.H. and Baral, N. 2018. 'Visitors' willingness to pay for snow leopard *Panthera uncia* conservation in the Annapurna Conservation Area, Nepal', *Oryx* 53.4: 633–42. https://doi.org/10.1017/S0030605317001636
3. Ibid.
4. Kanya, L. et al. 2019. 'The criterion validity of willingness to pay methods: a systematic review and meta-analysis of the evidence', *Social Science and Medicine* 232: 238–61. https://doi.org/10.1016/j.socscimed.2019.04.015
5. Schutgens, M., Hanson, J.H. and Baral, N. 2018. 'Visitors' willingness to pay for snow leopard *Panthera uncia* conservation in the Annapurna Conservation Area, Nepal', *Oryx* 53.4: 633–42. https://doi.org/10.1017/S0030605317001636
6. Hanson, J.H. et al. 2023. 'Assessing the potential of snow leopard tourism-related products and services in the Annapurna Conservation Area, Nepal', *Tourism Planning & Development* 20.6: 1182–202. https://doi.org/10.1080/21568316.2022.2122073
7. White, C. et al. 2015. *Cost–Benefit Analysis for the Reintroduction of Lynx to the UK: Main Report*. AECOM.

8. https://www.mayoclinic.org/diseases-conditions/lyme-disease/symptoms-causes/syc-20374651
9. https://aphascience.blog.gov.uk/2023/03/24/tb-day-2023/
10. https://www.gov.ie/en/publication/5986c-national-bovine-tb-statistics-2020/#
11. Bautista, C. et al. 2019. 'Large carnivore damage in Europe: analysis of compensation and prevention programs', *Biological Conservation* 235: 308–16. https://doi.org/10.1016/j.biocon.2019.04.019
12. https://www.swissinfo.ch/eng/society/protecting-livestock-from-predators-costs-swiss-taxpayers-millions/46345714
13. Smith, D. et al. 2016. *Reintroduction of the Eurasian Lynx to the United Kingdom: Results of a Public Survey*. Lynx UK Trust. http://insight.cumbria.ac.uk/id/eprint/3188/1/lynxinterimsurvey.pdf
14. Duffield, J., Patterson, D. and Neher, C.J. 2006. *Wolves and People in Yellowstone: Impacts on the Regional Economy*. University of Montana, Department of Mathematical Sciences.
15. Duffield, J. 2019. 'Shopping for wolves: using nonmarket valuation for informing conservation districts', in Rasker, R. (ed.) *People and Public Lands*. Bozeman, MT: Headwaters Economics.
16. Duffield, J., Patterson, D. and Neher, C.J. 2006. *Wolves and People in Yellowstone: Impacts on the Regional Economy*. University of Montana, Department of Mathematical Sciences.
17. Brazier, R.E. et al. 2020. River Otter Beaver Trial: Science and Evidence Report. https://www.exeter.ac.uk/research/creww/research/beavertrial/
18. https://environment.ec.europa.eu/topics/nature-and-biodiversity/habitats-directive/large-carnivores/eu-large-carnivore-platform/eu-funding-and-large-carnivores_en
19. https://commonslibrary.parliament.uk/research-briefings/cbp-9805/
20. Government of Ireland. 2022. *Budget 2023 Expenditure Report*. Dublin: Department of Public Expenditure and Reform. https://www.gov.ie/pdf/?file=https://assets.gov.ie/236053/e0ec55a5-f9bc-4d8c-b132-70560ca9fbe5.pdf#page=null
21. GFI, eftec and Rayment Consulting. 2021. *The Finance Gap for UK Nature*. https://www.greenfinanceinstitute.co.uk/wp-content/uploads/2021/10/The-Finance-Gap-for-UK-Nature-13102021.pdf
22. Sargent, N. 2019. 'Significant gap in biodiversity funding in Ireland', *Green News*. https://greennews.ie/significant-funding-gap-ireland/
23. O'Rourke, E. 2019. 'Drivers of land abandonment in the Irish uplands: a case study', *European Countryside* 11.2: 211–28. https://doi.org/10.2478/euco-2019-0011
24. https://www.geeksforgeeks.org/what-is-no-free-lunch-theorem/
25. Dickman, A.J., MacDonald, E.A. and Macdonald, D.W. 2011. 'A review of financial instruments to pay for predator conservation and encourage human–carnivore coexistence', *PNAS* 138.10: 13937–44. https://doi.org/10.1073/pnas.1012972108
26. https://sites.warnercnr.colostate.edu/centerforhumancarnivorecoexistence/wolf-conflict-reduction-fund/
27. https://www.gov.uk/government/publications/lynx-reintroduction-in-kielder-forest/lynx-reintroduction-in-kielder-forest-natural-england-advice-to-the-secretary-of-state#social-feasibility-and-risk-assessment

Chapter 14: For the Love of Wisdom

1. Preston, C. 2023. 'The spurious wild of wilderness', *The Philosophical Salon*. https://thephilosophicalsalon.com/the-spurious-wild-of-wilderness/
2. https://www.choosingtherapy.com/bilateral-stimulation/
3. Sandel, M.J. 2012. *What Money Can't Buy: The Moral Limits of Markets*. New York: Farrar, Straus and Giroux.
4. Adams, W.A. 2003. *Against Extinction: The Story of Conservation*. London: Earthscan.
5. Preston, C. 2023. 'The spurious wild of wilderness', *The Philosophical Salon*. https://thephilosophicalsalon.com/the-spurious-wild-of-wilderness/
6. https://dictionary.cambridge.org/dictionary/english/wisdom
7. https://wildlifenl.nl/en/
8. Drenthen, M. 2017. 'Environmental hermeneutics and the meaning of nature', in Gardiner, S.M. and Thompson, A. (eds). *The Oxford Handbook of Environmental Ethics*. Oxford: Oxford University Press. https://doi.org/10.1093/oxfordhb/9780199941339.013.15
9. Jones, P. and Pennick, N. 1995. *A History of Pagan Europe*. London & New York: Routledge.
10. Ibid.
11. White, L. 1967. 'The historical roots of our ecologic crisis', *Science* 155.3767: 1203–7. https://www.jstor.org/stable/1720120
12. Dean-Drummond, C. 2008. *Eco-Theology*. London: Darton, Longman & Todd.
13. McCarthy, P. 2003. *The Confession of St. Patrick and His Letter to the Soldiers of Coroticus*. Avoca: Parish of St Mary & St Patrick. https://www.loughderg.org/wp-content/uploads/2015/02/confession_of_st_patrick.pdf
14. Pope Francis. 2015. *Laudatio si': On Care for Our Common Home*. Encyclical letter. Dublin: Veritas Publications.
15. https://www.theguardian.com/world/2023/apr/27/dawn-of-the-new-pagans-everybodys-welcome-as-long-as-you-keep-your-clothes-on
16. https://www.statista.com/topics/4765/islam-in-the-united-kingdom-uk/
17. Hanson, J.H., Schutgens, M. and Leader-Williams, N. 2019. 'What factors best explain attitudes to snow leopards in the Nepal Himalayas?', *PLoS One* 14.10: e0223565. https://doi.org/10.1371/journal.pone.0223565
18. Waldau, P. *Animal Studies: An Introduction*. Oxford: Oxford University Press.
19. Massei, G. et al. 2010. 'Can translocations be used to mitigate human–wildlife conflicts?', *Wildlife Research* 37: 428–39. https://doi.org/10.1071/WR08179
20. https://uitspraken.rechtspraak.nl/details?id=ECLI:NL:CBB:2023:285
21. 1 Kings 4:29.

Chapter 15: Reconciliation

1. For a recent overview of this topic, see, for example, Burak, M.K. et al. 2023. 'Context matters when rewilding for climate change', *People and Nature* 6.2: 507–18. https://doi.org/10.1002/pan3.10609

Index

Aachen cathedral 129–30
Abruzzese Shepherd Dog 86
acoustic deterrence 91
Africa 23–4
Agivey graveyard, Aghadowey, Co. Derry/Londonderry 37–8
agriculture 26, 74–5 *see also* farmers and farming
AGRIDEA 89
agro-environmental systems 16
Ailwee caves, Co. Clare 5
Ale, Som 199
Almere, Netherlands 63
American Bear Foundation (ABF) 135
American bison (*Bison bison*) 150, 163–5, 167–9
American transcendentalists 40
Anderson Ramirez, Malou 91, 130–1, 134
Anderson Ranch, Montana 91, 123, 233
Annapurna Conservation Area (ACA) 47, 198–9
Annapurna region, Nepal 49, 98, 197–8
Antrim plateau, Northern Ireland 3–4, 6, 15
apex predators 68, 71, 76, 224–5, 235
apps 87
Arapaho National Wildlife Refuge, Colorado 150
archaeological excavations 37–8
archery hunting 67–8
Aughton, Mark 115

Bales, Jaden 68, 69, 70
Bali tiger 42
Ballybay Central School, Co. Monaghan 23, 136
Baral, Nabin 199
bears 30, 46, 135, 193 *see also* brown bear (*Ursus arctos*); Eurasian brown bear (*Ursus arctos arctos*); grizzly bear (*Ursus arctos horribilis*)
bear-proof bins 101, 134
bear spray 114–16
Beauty and the Beast (Disney) 23, 24, 27, 29–30
beavers 7, 59, 158, 173, 208
Belgrade, Montana 140
benchmarking and profitability 107
Bern Convention 1979 58, 194
Bible 227

Big Bad Wolf (Little Red Riding Hood) 25
BIJ12 (Dutch government agency) 97
biocentrism 228
biodiversity 227
birds of prey 7, 59
bison *see* American bison (*Bison bison*); European bison (*Bison bonasus*)
black bear (*Ursus americanus*) 43, 121, 135
Blantyre mission site, Malawi 101, 105
blue sheep (bharal) 49, 50–1
'Bob' (rancher) 151–3
Boone and Crockett Club 220
bounty hunting 105
Bozeman, Montana 142
Brexit 195, 241–2
Bringing Back the Beaver (Gow) 59
Britain: agricultural subsidies 210; attitudes to reintroductions 194; beaver 59; birds of prey 59; brown bear 4, 43; coexistence 77; costs of sheep losses 204–5; deer species 176; deterrence 95; ecological benefits 178; extinctions 4–5, 42–6; large carnivore reintroduction 159; lynx 4, 44–5, 59–60; rewilding 58–9; wolf 4, 45
British Association of Nature Conservationists 59
brown bear (*Ursus arctos*) 4
browsing patterns 177
buffalo *see* American bison (*Bison bison*)
Buffalo Ranch, Lamar Valley 165, 169
bullfighting 136
Bullmastiff 21
'the bullpen' 150
bureaucracy 14–15, 98, 111, 188, 191

camping 134, 150–1
Canadian lynx (*Lynx canadensis*) 113
capuchin monkey 182–3
cattle *see* livestock
cave hyena 42
cave lion 42
Center for Human–Carnivore Coexistence (CHCC) 109, 147, 154
certification schemes 143
charismatic animals 62
'Children's Guide to Animal Rights' 136
China 98
Chinese water deer 176

Christian perspectives 226–7
chronic fatigue syndrome (CFS) 169–71
Chur, Switzerland 82
civil society organisations 187
closed-canopy forests 39
Cody, Wyoming 113, 115, 117, 135, 140
coexistence: defining 71–3, 77; journey towards 74; models 31
coexistence funds 133
coexistence toolkit 221
coexisting with nature 240
collars (protection from predation) 88, 90
collective herd management 134
Colorado: compensation 104–7, 211, 219; Proposition 114 106, 109, 148–9, 153–4; reintroduction projects 147; wolf reintroduction 158
Colorado Parks and Wildlife (CPW) 147–8
Colorado State University (CSU) 108–9, 145, 147, 153–4
Colorado Wolf Conflict Reduction Group 154
Common Agricultural Policy 104
community-based compensation schemes 98
Community Benefit Society 187
compensation 88, 236; Colorado 104–7, 211, 219; and insurance 108; for livestock 98, 203, 205, 211; Netherlands 98–9, 104; safeguarding public funds 101, 109; secondary losses 106–7; as a subsidy 110; Switzerland 89, 104 *see also* financial tools
compost toilets 140, 143
conservation 11; and hunting 137; and injustice 36; and psychology 31–2; social processes 31, 36–7
Conservation Conflict Transformation 157–8
conservation grazing 65
conservation partnerships 6
conservation payments 108, 198
conservation translocation 124–5, 128
consulting communities and landowners 60, 111
contested spaces 221
contingent valuation 202
continuum of control 28
Convery, Ian 194
corrals 90
costs and benefits 103–4, 111, 203–9, 211–12
County Antrim 3–4, 10, 60
County Carlow 4, 45
County Clare 5
County Donegal 60, 240
County Down 53, 60
County Kerry 60
Covid-19 pandemic 187
coyote 119, 121, 152
Cromwell, Oliver 45–6, 190

Crone, Jack 35–6
crop damage 50
culling animals 66
Cumbria 81
cycling 216–17

Daltun, Eoghan 173, 174
dark stairs 23, 24
De Brazza's guenon 182
death 66
deer 173, 176–7, 206, 232
Dennis, Roy 59
Derry (Londonderry) 10, 37, 39
deterrence 85, 95
deterrence costs 205
deterrence methods 88–9, 91, 106, 123
Devon, England 208
Dickman, Amy 108, 133, 211
Disney 5, 23, 24, 173, 174, 177, 178, 239
DNA testing 97–8, 109
dog attacks 19–22, 25–6, 29
Doherty, Mitch 216
Drenthen, Martin 223–6
Dudh Khosi river, Nepal 48
Dungonnell reservoir, Co. Antrim 6
Dunsany Nature Reserve, Co. Meath 61
Dutton, John 122

eating animals 139
ecocentrism 228
ecological benefits 178
ecological changes 176–7
ecological economy 5–6
ecological renewal 173
ecological restoration *see* rewilding
ecology: and the economy 137; effects of lethal control 138–9; laws of nature 171–2; political contexts 179, 190, 196
economic perspectives 138–9, 200, 203–4, 238
ecosystem management 7, 68–9
educating visitors 133
electric cars 73–4
electric fences 89
elk (*Cervus canadensis*) 68, 152, 167, 172
Emerson, Ralph Waldo 40
emu 180, 181–5
England: beavers 158; 'A Green Future' environmental plan 195; lynx 44, 175; wolves 4, 175
enterprise methods 131–2
environmental hermeneutics 224
ethics 125–8
ethology 136
Eurasian brown bear (*Ursus arctos arctos*) 42–3, 58, 125, 193
Eurasian lynx (*Lynx lynx*) 4, 7, 42, 44, 58, 168, 175, 176

Eurasian grey wolf (*Canis lupus lupus*): attacking sheep 96–7, 100; attitudes to 99–100, 193; Cromwellian campaigns 45–6, 190; extinctions 4, 42–3, 46; habitat 175; hazed with paintballs 123; jumping electric fences 89; killing sheep and cattle 81; losing fear of people 124; population 58; prey 176; reintroduction of 168, 175
Europe 25, 42, 58
European bison (*Bison bonasus*) 7, 33–6, 59, 218
European Rewilding Network 58
European Union (EU): Common Agricultural Policy 104; Habitats Directive 1992 58, 194; LIFE programme 58, 209; Parliament 125
Evanston, Wyoming 119
Everest region 47–9, 98 *see also* Sagarmatha National Park (SNP)
extinctions 36, 42–3

fairy tales 5, 24–5
fallow deer 176, 206
farmers and farming 8–10, 13–15; for conservation 210; creating diverse agro-environmental systems 16; diversity 14; IFS perspectives 30–1; key to wildlife 12; part of landscapes 110–11; red kites predating lambs 54; reducing risks of carnivores 85, 235–6 *see also* agriculture
'farmer's ethic' 225
farming unions 193
fencing 88–9, 90, 94
Feral (Monbiot) 59, 139, 167, 218
financial incentives 109
financial losses 111
financial tools 104–5, 106, 108, 236 *see also* compensation
fladry 91
folk tales 25
food 175–6
Fort Collins, Colorado 145, 153
fox 9
Francis of Assisi 227
French Revolution 138
Friesian heifer 64
funding reintroductions 208–11

Gallatin National Forest, Montana 173
gaming the system 109
Garron plateau, Co. Antrim 3–4, 16, 243
Gaurishankar Conservation Area 50
Geismann, Stefan 82–3, 84, 89
Glacier National Park, Montana 216
'glassing' 67, 68

Glenariff, Co. Antrim 240
Glenarm valley, Co. Antrim 15–16
Glens of Antrim 5, 231, 232
Global South 40
goats 83, 232
golden eagle 7, 60
Gonzalez, Mireille (Ray) 146, 147–9, 153–4, 157
Gotland, Sweden 38, 150
governance 73–4, 146–7, 155, 155–9, 237
Governing the Commons (Ostrom) 156
government 185–9
Gow, Derek 59
GPS 126
grants 109
Graubünden, Switzerland 81, 94, 125–6
Greater Vancouver, British Columbia 33–4
Greater Yellowstone Ecosystem 131
grizzly bear (*Ursus arctos horribilis*): attacks 93; biology 92, 93; hunting 137, 139; licenses to cull 121; Montana 130; Tom Miner Basin 91, 135
grizzly-proof bear bins 134
grizzly tourism 133–4
The Guardian 172

Haast's eagle 42
habitats 175
Half-Earth movement 110, 243
Halloween 213–14
Hanson, Amanda 150
Hanson Park, Walden, Colorado 150
Hanson, Paula 169–71, 215
Hardin, Garrett 156
harvesting predators 138–9
hazing 123–4, 127
Heck cattle 65–6
Henderson, Henry 101
herbivore herds 42
hermeneutics 224
Hickey, Kieran 4
High Nature Value farming 210
Himalayas 197–8
history 36–7, 38–9, 56
history of rewilding and reintroductions 50, 52, 235
history of the wild 40
Hometree, Co. Clare 61
horses 232
household surveys 198–201
human deaths by dogs 215
human health 204
humans (*Homo sapiens*): absence of 40; apex predators 68, 71, 235; dumb things 133; migrations 23; ultimate predators 90
human-induced extinctions 41
human nature 27–32, 154, 229

human–wildlife coexistence 62, 72–4
hunting 67–71, 132, 137–40, 143
husbandry 85, 110

identity politics 157
Imperial Circus, Rome 43
innovation 188
Inside Out movies 29
insurance 107, 236
intergenerational trauma 25–7
Internal Family Systems (IFS) 28–31, 41, 55
Ireland 42–3; agricultural subsidies 210; brown bear 4–5; coexistence 77; deer 176; deterrence 95; ecological benefits 178; habitat suitability 175; large carnivore reintroductions 60, 159; lynx 4–5, 44, 173, 175–6; National Biodiversity Action Plan 194–5; National Parks and Wildlife Service 210; raptor reintroductions 60; rewilding 58–9, 61; wolf 4, 5, 45–6, 239–40
Irish brown bear 43
Irish Examiner 173
islands 42 *see also* theory of island biogeography
issues of life and death 117–18

Jackson County, Colorado 150, 151, 153
Japanese Akita 181
Japanese macaque 182
Jubilee Farm, Co. Antrim 9–10, 12, 74, 83, 89, 185–7, 227
Junction Butte wolf pack 172

Kemmerer, Wyoming 121
Kent, England 7, 59, 218
Kielder Forest, Northumbria 7, 59–60, 175, 204–5, 206, 207
Kilgreany cave, Co. Waterford 44
Kipling, Rudyard 189, 191
Knepp Castle Estate 59
Kondelis, Joe 135, 137–40
Konik pony 65–6

Lama, Rinzin 199, 200
Lamar valley, Yellowstone 58, 172, 218
Lambert, Joanna 163–9, 172–3, 174–5, 176–7, 178, 218
lambs 54
land abandonment 210
land sharing 75
Lander, Wyoming 69, 114
landscapes of fear 69, 167, 174–9, 237
Langwies, Switzerland 88
large carnivores: costs 156; ecological saviour status 177; evoking powerful emotions 157; fairy tales vilifying 25; human-induced extinctions 41; IFS perspectives 30, 31; intergenerational trauma 26–7; living alongside 81; as a public good 156–7; relating to 22; roaming range 124, 155–6; translocating 124
large carnivore conservation 71, 135–7
large carnivore reintroductions 7–8; attitudes to 193–4; Britain and Ireland 60, 188, 195–6, 243–4; complexities of managing and governing 242; correcting historical wrongs 166; costs 204, 208, 212; economic case 201; failure and mortality rates 178, 228; philosophy 217; planning processes 153; polarised perspectives 57; political ecology 191; reciprocity 174; role of government 188, 189; social processes 36–7, 245; trial-and-error nature 211; Yellowstone wolves as a benchmark 207
'The Law of the Jungle' (Kipling) 189
laws of nature 177–8
legal considerations 194
Lelystad, Netherlands 63
lethal control 125–8, 127
Leyen, Ursula von der 125
livestock 84–5, 103–4
Livestock Board 141
livestock compensation 98, 203, 205, 211 *see also* compensation
livestock losses 85, 104, 232
livestock-protection dogs 85–8, 90, 119
Londonderry (Derry) 10, 37, 39
Londonderry, Frances Anne Vane Tempest, Lady 4
long-term public funding 209–10
Longstanton, Cambridgeshire 20–1
Lukla, Nepal 48–9
Lyman ranch, Wyoming 119–21
Lyme disease 204
lynx 30, 43, 168, 175–6 *see also* Canadian lynx, Eurasian lynx
lynx reintroductions 59–60, 173, 178, 192–6, 204–5
Lynx to Scotland project 7, 60
Lynx UK Trust 7, 59–60, 139, 204–9, 211

malaria 101–5
Malawi 10–11, 101–3, 105, 117, 119–20, 221
Malawi Mountain˙ Gin 68
Mammoth Hot Springs 172, 174
management methods 127–8
Marcus, Hugh 12–16, 232–4, 239
marginal farmers 191, 196
Markermeer, Netherlands 63
market research 198
Martial (Roman poet) 43

Maslow's hierarchy of needs 166–7, 171
mathematics and machine learning 210
meat consumption 136
meso-predators 177
Missing Lynx Project 7, 60
Missoula, Montana 213–14, 216, 221–2, 229
Mollie's pack 172
Monbiot, George 59, 127, 139, 167, 172, 174, 218
monkey sanctuary 180–4
Montana 122, 129, 130
Montana Grizzlies 219
moose (*Alces alces*) 115, 137, 152
moral experiences 224–8
moral hazards 110
mosquitoes 102–3, 105, 109
mountain lion 117, 119, 121, 152
Mountain West landscapes 118–19
Mourne, Co. Down 53–4
Mugs Old Town, Fort Collins 147
multicultural societies 227
multipurpose rural landscapes 111
multi-stakeholder governance 158
muntjac deer 176, 206
myalgic encephalomyelitis (ME) 169–71

Nabin Baral 109
Namche Bazaar, Nepal 48–9
National Geographic 134
Native Americans 221
natural selection 171–2
nature 226, 229, 237
nature recovery 195 *see also* rewilding
Nelson, Gaylord 190
Nepal 11–12, 27, 47, 50–1, 85, 198, 203
Netherlands 28, 193, 223; compensation 98–9, 104; DNA testing 98, 109; non-lethal projectiles 124; wolf 123
neutral facilitators 109
Nietzsche, Friedrich 216
night-time corrals 90, 100
Nijmegen, Netherlands 223
Nissan Leaf 74
'No Free Lunch' theorem 210
'noble savage' (Rousseau) 40
non-lethal projectiles 124
non-native species 176–7
'non-use' values 132, 201
North America 54, 56
North American grey wolf (*Canis lupus*): Colorado 151–2; decline in the USA 165–6; and ecological renewal 173; habitat 175; reintroducing to Yellowstone 57–8, 147–8, 167; shooting 117; Wyoming 121–2
Northern Ireland 9–10, 53–4, 149, 241

Northern Ireland Science Festival 220, 243
Norway 204–5, 208
Nuffield Farming Scholarships Trust 12

oikos ('home') 5, 137, 241
Oostvaardersplassen Nature Reserve 58, 63–6, 223
Ostrom, Elinor 156

paintballs 123, 124
Papi, Andrea 125
'Payments to Encourage Coexistence' fund (Dickman) 108, 211
peace and reconciliation studies 148, 154, 157
peer-to-peer support networks 94
penalties and incentives 103
Peruvian shepherds 93, 120
philosophical approaches 213–30; large carnivore reintroductions 217; love of wisdom 222–3; predators in permanent captivity 125; vision and values 230; wisdom 222
Pleistocene defaunation 42
Pocahontas (Disney) 239
polar bear 43
political capital 241–2
political ecology 190–1, 196, 205–6
politics 188–9
PPE+PE acronym 32, 190
pre-Christian perspectives 226
predation 232
predation sequence 88
'predator-friendly' certification 140–2, 143, 198
predator tourism 108
predators 234; actual and perceived risks 84–5; the 'commons' 156; competing with 23, 95; ecological benefits 70–1, 138–9; extinctions 42; focusing on weaker individuals 69; in permanent captivity 125; polarised perspectives 57; as threats 23; tolerance 12; troublemaking 125
Preston, Christopher 213–23
prevention 101, 104, 106, 110
proactive payments 107–8, 236
productive terrestrial ecosystems 172
profitability and benchmarking 107
Proposition 114 106, 109, 148–9, 153–4
protected statuses 195
psychic multiplicity 28, 55, 157, 189, 222
psychology 27, 31–2
public money for public goods 244
Putten, Gelderland, Netherlands 96
Pyrenean Mountain Dog 86, 119

Queen's University Belfast 38–9, 60, 136, 213, 214, 243
Queen's University Management School 10, 136

radio collars 126
ranchers 152, 153
range riders 92–3, 130, 134
raptors 7, 59
recreating the past 177
red deer 65–6, 176, 206
red kite 7, 53–4, 60
referenda 125 *see also* Proposition 114
regenerative economy 143–4
'Renewed Coexistence' 158
reproduction rates 178
Restoring the Wild (Dennis) 59
rewilding 50–67; competing visions 56; as cultural landscapes 229; defining 6–7, 55; ecosystem management 68–9; history of 51–2; Ireland 61; as a loaded term 57; mixed feelings 31–2; North America 54, 56; as a process 61; social processes 36–7, 245; what, where and when 55–6; and the wild 36
Rewilding Britain 60
Rewilding Europe 58
rights of animals and people 137–8
Robobear 113, 115–16, 117
roe deer 176, 206
Rogers, John 134
Romanticism 40–1
Roosevelt, Theodore 185
Rose Creek, Lamar valley 165
Rose Creek pen 164, 166, 167
Rousseau, Jean-Jacques 40
rubber bullets 124
rugby 220
rural and urban communities 192–3, 194, 205
rutting 168

sabretooth cat 42
Sagarmatha National Park (SNP) 47–8, 49, 198
Sami reindeer herders 107–8
Sandel, Michael 219
Savings and Credit Cooperatives 98
Schumanns, Jorien 199, 200
Schutgens, Maurice 198–9
science communications 173
Scotland 43; beaver 7; lynx 7, 44, 175; multi-stakeholder governance 158; support for reintroductions 192; white-tailed eagle 158; wolf 4, 45, 175
Scottish Highlands 175
Scottish Rewilding Alliance 192, 196

Scottish Southern Uplands 175
sea eagle 60
secondary losses 106–7
'self-willed nature' 225
Selway-Bitterroot Wilderness, USA 216
semi-wild places 71
sentimentality 127
sheep 96–7, 204–5, 232–3
shepherds 88, 93
shooting animals 117, 121, 122
Shrestha, Niki 199
Siberian tiger 123
sika deer 176
Sims, Shaun 119, 120–2, 124, 190, 233
Slemish, Co. Antrim 13
smaller carnivore species 177
snow leopard 27, 47, 49, 85
Snow Leopard Conservancy 11–12, 98
snow leopard conservation 11–12, 90, 107, 198–9
Snow Leopard Conservation Action Plan 202
'snow leopard trail' 202
snowstorm March 2013 14
social enterprises 185
social processes 36–7, 39, 245
social sciences 37
Sonia (wolf) 45
Spain 136
spatial issues 42, 178, 237–8
'speak softly and carry a big stick' (saying) 185, 189
St Andrews International High School 102
St Patrick 13, 227
storytelling 24–5, 209
subsidies 110, 210
surveys 199–201
Sweden 107–8
Swiss Farmers' Union 81, 82, 94
Switzerland 81–4, 94, 193; costs of coexistence 103–4; lethal control legislation 125–6; livestock losses 85; non-lethal projectiles 124; shepherd huts 93; wolf attacks 90

targeted translocations 124
Tasmanian wolf (thylacine) 42
theory of island biogeography 42, 59, 178
Theory of Island Biogeography (Wilson) 75
Thetford Forest, Norfolk–Suffolk border 7, 59–60, 175, 204–5, 206, 207
Thirteen Mile Lamb and Wool Co 140
Thompson, Dan 114, 122, 125, 126, 127
Thoreau, Henry David 40
Tom Miner Basin Association (TMBA) 133–4
Tom Miner Basin, Montana 91–3, 129, 132, 133, 135, 233

tourism 49, 130–5, 138, 143, 202, 207, 210
'The Tragedy of the Commons' (Hardin) 156
transhumance 122, 233
translocations 116–17, 121, 122, 124–5, 127–8, 177
trapping 122
trauma 22, 25–7
Tree, Isabella 59, 134
trial-and-error nature 211
trophic cascades 54, 167, 174, 175, 176, 189, 237
troublemaking predators 125
Twain, Mark 158
Tyler, Dave 140, 142, 143

Ulster *see* Northern Ireland
Ulster Way 15–16
United Kingdom (UK) 194, 210; Dangerous Wild Animals Act 1976 194; devolved administrations 194; Environment Act 2021 194
United States of America (US) 135, 165–6, 193; Endangered Species Act 1973 57; Forestry Service 134
University College Gotland 38
University of Cambridge 11
University of Montana (UM) 214
Untervaz, Switzerland 82, 83, 89
upland farming 94
Upper Marshyangdi valley, Himalaya 197–8

valuing the intangible 201
Vane Tempest, Frances Anne, Lady Londonderry 4
Västergarn, Gotland, Sweden 38
vegetarianism 136
Vikings 25
Vincent, Ellery 91, 92–3, 130, 133
Vital Ground 216

Walden, Colorado 150–1
Wales 175
walking 216–17
Weald of Kent 7
Weed, Becky 140, 141–3
werewolves 45–6
Wetering, Richard van de 96–7, 99–101, 190
Wetering, Stefana van de 96–7, 99–101, 190
White, Lynn 227
white-tailed sea eagle 7, 158
wild: defining reintroductions 216; and the domestic 28, 75; as non-human nature 75; philosophical concept 222; and a place 41; and rewilding 28, 36; and Romanticism 40–1; as uninhabited primeval forest 40

wild cat 12
Wild Nephin project, Co. Mayo 61
wild spaces and wild species 54–5
Wilder Blean project 59
Wilding (Tree) 59, 134
Wildland Research Institute 59
Wildlife Friendly Enterprise Network (WFEN) 143
wildlife populations 66
wildlife tourists *see* tourism
WildlifeNL 223–4, 225
'Willingness To Pay' (WTP) 201–2, 206, 208
Wilson, E.O. 75
wisent *see* European bison (*Bison bonasus*)
wolf 30, 43, 99–100, 175 *see also* Eurasian grey wolf (*Canis lupus lupus*); North American grey wolf (*Canis lupus*)
Wolf Conflict Reduction Fund (WCRF) 108–9, 154, 211
Wolf Conflict Reduction Group 158
wolf reintroductions 174–5; Colorado 157; holding pen 164–5; legal considerations 194; support for 192–3; Yellowstone 57–8, 147–8
wolf-related spending 131
wolf risk zones 98
'wolves-change-rivers' 56, 172, 174, 218, 237
Wolves in Ireland (Hickey) 4
wood-pasture mosaics 39
wool producers 142
written histories 25
Wyoming 93; archery hunting season 67–8; managing coexistence 126; moose hunts 137; threatening bears 116–17; translocations 125; as an unscrupulous place 122; wolves 121
Wyoming Wildlife Federation 68, 70, 71
Wyoming's Game and Fish Department (WGFD) 113–15, 126

XL Bully dog 19–21, 22, 25–6, 215

Yellowstone Institute 165
Yellowstone National Park 28, 40, 113–14, 163–8; ecological benefits 178; reintroducing wolves 57–8, 147–8, 167; tourism 207; traffic congestion 132–3; wolf 117, 130
Yellowstone (TV series) 122
YouGov polls 192

Zimbabwe 123
Zippert, David 85–7
Zippert, Sarah 81–2, 87, 89, 190
Zuid-Kennemerland National Park 35, 35–6, 40, 63, 223, 224